Introduction to Molecular Science

INTRODUCTION TO MOLECULAR SCIENCE

Edited by:

Ageetha Vanamudan

Delve
PUBLISHING

www.delvepublishing.com

Introduction to Molecular Science

Ageetha Vanamudan

Delve Publishing
224 Shoreacres Road
Burlington, ON L7L 2H2
Canada
www.delvepublishing.com
Email: orders@arclereducation.com

ISBN: 978-1-77469-429-9 (Hardcover)

ABOUT THE EDITOR

Dr. Ageetha is presently working as a Deputy Registrar at Team Lease Skills University, Vadodara. She Did her M.Sc & M.Phil in Chemistry from Pondicherry University. She completed her PhD in Analytical Chemistry from M.S.University, Vadodara. Her research interest is on Nanomaterials, catalysis and Environmental chemistry. To add to her credits, she has published 14 articles in various International Journals. She has 12 years of teaching experience. Also she serves as a consultant to various industries.

TABLE OF CONTENTS

LIST OF FIGURES

ABSTRACT

Molecular science is a field that explains the mechanisms at work in the diverse and rich molecular world and contributes to the development of new technologies to address the energy and environmental problems necessary to support a sustainable society. Molecular science primarily involves working in laboratories and examining various human samples, including fetal cells, hair follicles, blood, and bones. This volume is useful in understanding the basis of molecules and chemistry.

Chemistry and biochemistry are developing rapidly as disciplines. Chemists and biochemists are increasingly involved in work aimed at solving larger societal issues such as energy, sustainability and human health. Making progress in these areas requires a solid understanding of chemistry and biochemistry, but often also learning to work in interdisciplinary teams or work on projects that go beyond the boundaries of traditional chemistry and biochemistry. Molecular science provides an in-depth understanding of fundamental principles, as well as the broader skills and experience needed to address a variety of environmental problems.

PREFACE

Molecular science is a field that studies the chemical structures and processes of environmental phenomena involving molecules, the basic unit of life. Its branches include molecular biology, which specializes in the study of nucleic acids, proteins, and large molecules essential to life processes and also to know how these molecules interact and behave in the cells. Molecular biology fundamentally arose in the 1930s from closely related disciplines like biochemistry, genetics, and biophysics; it still shares many similarities with these disciplines today.

Various techniques have been developed for molecular sciences. In particular, molecular science attempts to understand the three-dimensional structure of macromolecules through techniques like X-ray diffraction and electron microscopy.

In particular, the discipline seeks to understand the molecular basis of genetic processes. Molecular science can be used to map the location of genes on specific chromosomes, assign those genes to specific traits of an organism, and use genetic engineering (recombinant DNA technology) to isolate, sequence and modify specific genes. These methods may also include techniques such as polymerase chain reaction, Western blotting, and microarray analysis.

Discovery at the molecular level is indispensable in almost all fields of study in the natural sciences today. Molecular science forms the basis of these different fields of study, and the interplay between them is strengthened and extended to a wider range of fields including primary physics, astronomy, information technology, environmental science, life sciences, pharmacy, and medicine.

The types of molecules to be studied are growing quite rapidly, showing a wide variety of changes in size and shape, from exotic molecules found in interstellar space to giant molecules such as proteins composed of more than 10,000 atoms, and molecules that regularly form supramolecular systems. Presently, molecular science continues to explain the mechanisms at work in the diverse and rich molecular world, contributing to the development of new technologies to address energy and environmental issues critical to supporting sustainable societies.

Most of the matter around us, including our bodies themselves, consist of molecules, which are truly microscopic matter, measuring 1/100,000,000th of a millimetre in size. The liquidity of water flow arises from the secret harmonious random movement of a large number of molecules (over 10 billion × 10 billion water in a single drop).

The distribution of the molecules varies from our local area to colossal scales in the universe. The bearing of particles controls the worldwide climate of the earth. For example, ozone particles in the stratosphere block unsafe UV light from the sun. There exist colossal amounts of atoms in the environment, which provide the foundation for life on the planet. The basis of the atoms and molecules are discussed in the book.

CHAPTER 1

THE NATURE OF CHEMISTRY

CONTENTS

1.1 INTRODUCTION

Substances (particularly referred to as elements or compounds) and their interactions are studied in chemistry, which focuses on their properties, composition, and structure. At least one element can be found in all substances, whether they are generated naturally or by synthetic method. Although they are made up of smaller particles, atoms are the basic building blocks of chemical compounds. It's impossible to have a lower quantity of anything than one atom of oxygen, mercury, or gold, for example. Thus, chemistry deals with the properties of individual elements and their interactions with one other, as well as how this knowledge may be used to achieve certain goals.

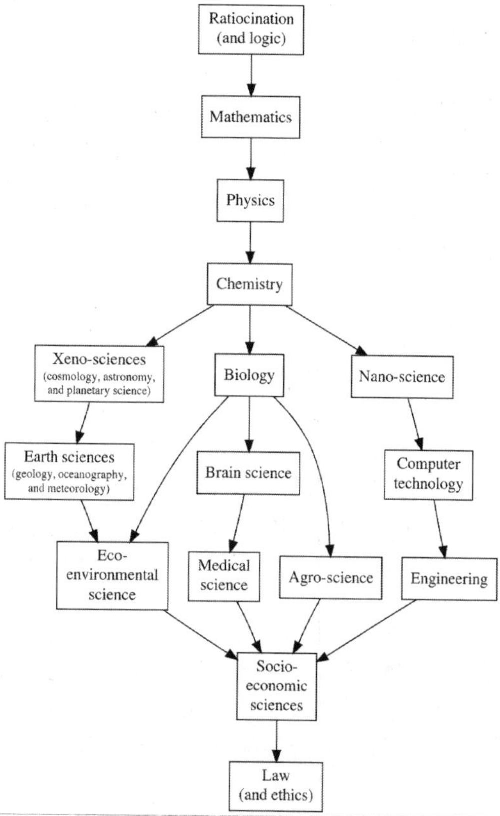

Figure 1.1: The Nature of Chemistry as the central science.

Source: By M stone, CC BY-SA 3.0, https://commons.wikimedia.org/w/index.php?curid=3016408

The development of a cohesive explanation for the Complex behavior of materials is a key task in chemistry. An explanation for why materials appear as they do, what gives them their permanent qualities, and how interactions between various substances may lead to the development of new substances and the disintegration of old ones are all part of this discipline. Chemistry has been working on theories of matter and there have been attempts to logically describe the material universe (Vasapollo et al., 2011). These concepts have sparked a lot of discussions. Atomic and molecular structures that can't be destroyed can change, while structures that can't be destroyed can't change. This is the difference between theories of change and those that claim that structures that can't be destroyed can't change.

Therefore, ideas of change are based on the capacity of unbreakable atoms or molecules to rearrange their configurations, whereas theories of permanence are founded on the ability of unbreakable chemical reactions that are the outcome of a wide range of interactions between different substances and their atoms and molecules, as well as their atomic structure and chemical composition. Chemists are interested in both synthesizing new compounds and repurposing already existing ones. Several chemical processes have been around since the beginning of time, such as cooking, fermentation, the making of glass, and metalworking.

Chemical technology includes vinyl, teflon, liquid crystals, semiconductors, and superconductors. Humans made significant progress in their understanding of the chemistry of living organisms over the twentieth century. Health and sickness can be better understood from a molecular perspective. DNA (deoxyribonucleic acid), which is composed of hundreds of millions of atoms, may be studied using contemporary chemistry's ever-increasingly advanced approaches to study subatomic particles. Compounds might be developed even before they are made to improve their properties. It's incredible how quickly new chemical knowledge is being collected (Turney et al., 2012). Humans have discovered and synthesized more than eight million distinct chemical compounds over human history. These substances can be found in nature or produced artificially.

1.2 TYPES OF NATURE OF CHEMISTRY

1.2.1 Analytical Chemistry

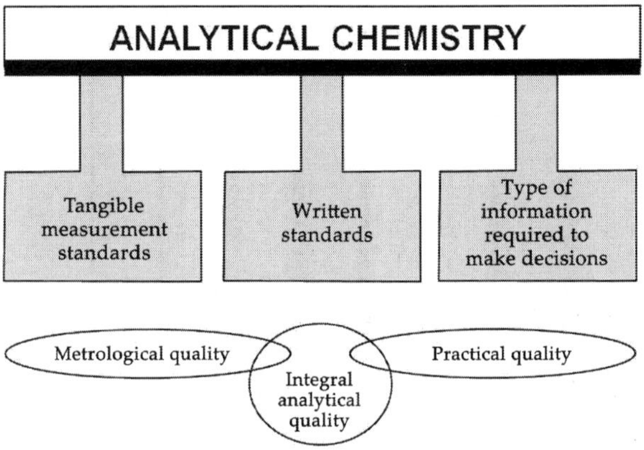

Figure 1.2: Analytical chemistry.

Source: *https://www.intechopen.com/media/chapter/40712/media/image4.jpeg*

Chemical compounds make up the vast bulk of Earth's elements. Wood, coal, minerals, and air are all examples of these elements. Every single pure chemical substance (such as oxygen or iron) has a unique set of characteristics that distinguish it distinct from the rest. When iron is exposed to oxygen, it forms hematite and magnetite, both of which have high magnetic properties. Iron has a silvery-white color, a melting temperature of 1,535 degrees Celsius, and is exceedingly ductile. Qualitative analysis is one of the subfields of analytical chemistry. For example, magnetite may be used to identify iron in different metal mixtures and compounds (Thomford et al., 2018). It is known as quantitative analysis when a specific amount of a chemical can be determined from a mixture or compound. Quantitative analysis shows that iron constitutes 72.3% of the total mass of magnetite.

In nature, magnetite may be found on beaches and riverbanks because it looks like black sand. A chemical process can produce distinguishable products. Thanks to the efforts of chemists. Over time, these products can be utilized to identify the presence of certain elements. Iron content in a sample larger than or equal to 1 ppm can be found using chemical methods. Qualitative tests can detect extremely small concentrations of certain chemical elements. Even if the item being burnt contains just one billionth of a gram of sodium, it will produce a yellow hue. This is due to the yellow hue that sodium emits when burnt. A wide range of impurities, as well as the properties of ultrapure substances, may be evaluated using analytical methods developed and employed by chemists. Impurity levels of less than 0.1 percent are common in laboratory chemicals, which are commonly employed in research. To meet particular standards, compounds with impurities lower than 0.001 percent can be obtained. Each branch of chemistry benefits from studying chemical combinations and naming the pure chemicals that make them up.

Analytical chemistry has never been more important in history than it is today. The current need for a broad range of nutritious meals, low-cost consumer goods, abundant energy, and technological breakthroughs that save time and effort is putting a huge strain on the environment. Chemical processes produce waste in addition to the valuable molecules they produce, and this waste is not always disposed of suitably. The environment has been harmed since the dawn of civilization, and the issues brought on by pollution have only become worse as the world's population has risen (Shin et al., 2019). Environmental protection calls for the widespread application of analytical chemistry procedures. These substances must be identified, the source of their origins established, and safe neutralization or removal processes must be created as quickly as feasible. A pollutant is considered dangerous to human health if its concentration exceeds a certain threshold. The next stage is to find toxic compounds at concentrations that are far lower than the danger level. Analytical chemists have a professional obligation to produce more precise and sensitive procedures and equipment.

Computer-aided analytic tools are making it easier than ever before for chemists to identify a wide range of chemical compounds. An absorbent but porous material is passed through an extremely long and thin column to separate the elements of gaseous mixtures using gas chromatography. Different gases interact with this absorbent substance in different ways, and they flow through the column at different rates. A mass spectrometer may be used to classify chemicals based on the mass of the ions that make them up when the gases exit the column. As little as one or two parts per billion can be detected by utilizing a mass spectrometer and a gas chromatograph to identify a chemical mixture's elements. A technique like atomic absorption or polarography or neutron activation, if used properly, might provide a degree of sensitivity comparable to or even greater than that of other methods. Most analytical equipment is made obsolete within a decade after its debut because of the rapid rate at which new technologies are created.

Environmental chemists and pharmaceutical chemists both benefit from modern technology since it is more accurate and faster.

1.2.2 Inorganic Chemistry

Figure 1.3: Inorganic chemistry.

Source: By Benjah-bmm27 - Own work, Public Domain, https://commons.wikimedia.org/w/index.php?curid=1010415

Modern chemistry was born in the late 18th century when chemists discovered the notion of the conservation of mass. Modern chemistry began by concentrating on substances that had little biological significance. Inorganic chemistry is concerned with the study of substances that contain very little or none at all, such as elements and compounds. Early on, the major goal was to identify the building blocks of all more Complex entities, such as atoms, molecules, and ions (Shi et al., 2010). Although elements like gold and carbon have been known since antiquity, many more were discovered and studied in the late nineteenth and early twentieth centuries. More than a hundred compounds have been located thus far.

Some of the most fundamental notions in contemporary chemistry, such as the rule of definite proportions, have been derived from the study of simple inorganic compounds like sodium chloride (common salt). This law states that the mass proportions of the constituent elements in the vast majority of pure chemical compounds are always consistent. The sodium content of salt is 39.3 grams per 100 grams, while the chlorine content is 60.7 grams per 100 grams.) Crystalline salt, which is known as halite, is formed when sodium and chlorine are mixed in equal amounts, producing one sodium atom for every one of chlorine. Two distinct elements are required for the formation of a binary compound. Although binary compounds have a limited structural variety, they are prevalent in inorganic chemistry. As a result, despite the enormous number of elements that might undergo reactions, the number of inorganic compounds is restricted. When three or more elements are included in a material's composition, the structural possibilities of that material are greatly expanded.

During the early half of the twentieth century, inorganic chemistry experienced a period of relative stagnation. A paradigm change in our understanding of the design of inorganic molecules has been brought about by the unique structural properties of boranes, which are compounds containing boron and hydrogen. Inorganic polymers may also be formed from certain inorganic substances, a discovery that was previously thought to be restricted to carbon molecules. Elements of ceramics are made of inorganic elements, as well as oxygen (Antoni et al., 2015). For years, ceramic artifacts have been created by repeatedly heating a jar of a mineral powder mixture. Ceramics are prone to breakage, although they are quite tough and can tolerate extremely high temperatures. Researchers are now working to produce new ceramics with the strength required to serve as turbine blades in jet engines. Internal combustion engine elements may eventually be made entirely out of ceramics rather than steel.

The existence of superconductivity was discovered in 1987 when researchers discovered that an alloy with the approximate formula YBa2Cu3O7 was superconducting at a temperature of about 100 K.. This new ceramic might likely be used in several electrical and magnetic applications because superconductors don't impede the flow of electricity. Because it's so simple, a high school chemistry class student could make an ultra-thin ceramic with superconductivity. To exemplify the unpredictable nature of the chemical sciences, the discovery in question shows how fundamental discoveries may still be discovered using low-cost chemicals and basic apparatus.

Recent advances in inorganic chemistry have led to some of the most fascinating discoveries. Thanks to collaborative efforts across other fields of study. Organometallic chemistry is a branch of chemistry that studies molecules that are both inorganic and carbon-heavy. Many different organometallic compounds are used as catalysts in industrial chemistry. Molecules that have the power to accelerate a process even when only a small amount is present are known as catalysts. These catalysts have shown some potential in converting natural gas into chemical compounds that are chemically similar to the original but significantly more lucrative. Researchers have also created enormous inorganic molecules that have a nucleus of metal atoms, such as platinum, surrounded by a shell of a variety of chemical elements (Sneskov & Christiansen, 2012). Two distinct kinds of molecules were combined to generate new compounds.

While some of the "metal clusters" exhibit properties that are akin to metals, others exhibit physiological effects. Biological systems such as respiration, neuronal activity, and cellular metabolism all rely on extremely small amounts of metals. In the discipline of bioinorganic chemistry, Even though organic molecules were traditionally regarded to be defining the chemical attribute of living creatures, inorganic chemistry today plays an important role in the scientific community.

1.2.3 Organic Chemistry

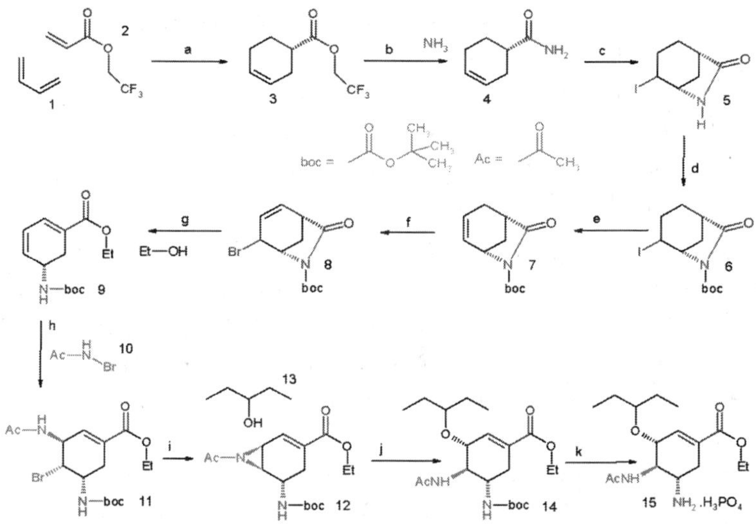

Figure 1.4: Organic chemistry.

Source: By User V8rik on en.wikipedia - Originally from en.wikipedia; description page is (was) here16:37, 28 April 2006 V8rik 1043x745 (18,952 bytes) (Corey oseltamivir synthesis), CC BY-SA 3.0, https://commons.wikimedia.org/w/index.php?curid=812699

Inorganic compounds are the result of the chemistry of carbon. When carbon atoms come into contact with one another, they may build an astonishingly wide variety of three-dimensional structures. Photosynthesis, a chemical reaction involving carbon dioxide and water, yields oxygen and carbohydrates. Polymeric carbohydrates include cellulose, which gives plants their structural stiffness, and starch, which plants employ to store energy. In the world of plants and animals photosynthesis produces a wide variety of organic compounds, including simple sugars. As the amounts of carbon, hydrogen, oxygen, nitrogen, sulfur, and phosphorus in the resultant molecules may be varied to any extent, the number of molecules produced much exceeds the total number of inorganic molecules (Stuppia et al., 2012). These naturally occurring molecules are the core focus of organic chemistry's research and development efforts.

Many naturally occurring chemicals share a small number of building units. Oxalic acid and formic acid are also present in rhubarb leaves (HCO_2H). C_2H_5OH, the alcohol derived from fermenting fruit, contains the ethyl ($C_2H_2O_4$) group. Penicillin, vitamin B12, proteins, and nucleic acids are examples of Complex natural products. Natural products are present in very small proportions, and so are notoriously difficult to separate from the organisms that produce them. If the sample is isolated in the purest form possible, sophisticated experimental approaches can provide structural information even for a millionth of a gram of material.

Understanding the interplay between molecular structure, physical and chemical characteristics is the goal of physical organic chemistry. It is possible to synthesize new compounds with the desired properties if the structural elements known as functional groups have been identified as the basis of a substance's distinctive properties and traits. Synthetic chemistry refers to the controlled production of certain compounds in a laboratory setting. For certain substances, manufacturing them rather than extracting and purifying them from natural sources is easier. Vitamin C, which is produced in large quantities each year, is one such example. Synthetic drugs come in a wide variety of forms, each with its own set of peculiarities that add to the intrigue. Plastics, as well as several pharmacological and agricultural substances, are notable examples.

Due to the intricacy of organic substances, synthetic chemists are usually greeted with a never-ending source of disappointment.

There are precise steps and interactions in three dimensions that need to be followed while trying to synthesize a specific molecule. To create distinct compounds, a specific number of atoms can be brought together in a variety of ways. Similar to how a particular amount of bricks and wood may be used in several ways to build a variety of different buildings. Although there are an endless number of possible structural configurations, only one of them can be found in nature. There are 37 carbon and 67 hydrogen atoms, 13 oxygen and one nitrogen. This is the basic structure of the drug. Although these 118 atoms may build 262,144 distinct structures, only one of these forms shows the features of natural erythromycin (Rhoads et al., 2012). In addition to their role in the chemistry of living organisms, organic molecules are exceedingly complex and exciting to study because of their wide range of structural compositions. Of all the branches of chemistry, organic chemistry possesses by far the most data of them all.

1.3 BIOCHEMISTRY

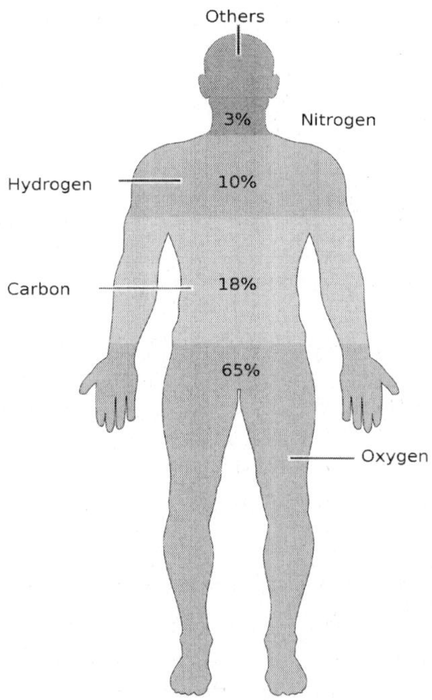

Figure 1.5: Biochemistry.

Source: By OpenStax College - This is a truncated version of File:201 Elements of the Human Body-01.jpg, CC BY 3.0, https://commons.wikimedia.org/w/index.php?curid=46182835

A growing body of inanimate chemistry knowledge-pushed scientists in the nineteenth century to explore the molecular structure and reactivity of live organisms in an attempt to define their physiological activities. A new scientific field was born out of these efforts: biochemistry. Biochemists employ the techniques and concepts of chemists to investigate the chemical bases of life. In general, it is acknowledged that an organism's physiology is the result of tens of thousands of chemical interactions carefully coordinated with one another. Biochemists have established the principles of energy transfer inside cells, cell membrane chemical structure, genetic information coding and transmission, muscle and nerve function, and biosynthetic mechanisms (Rastogi & Sani, 2011). Animals as diverse as bacteria and as familiar as humans share comparable macromolecules that perform analogous functions. It's far more difficult to do biomolecular research because of the numerous roadblocks.

In general, these molecules are large and have a lot of structural complexity. The chemical reactions in which these molecules are involved also happen very quickly. To provide one example, it takes a millionth of a second to break two DNA strands apart. Biomolecules known as enzymes provide an intermediate function that allows the reaction to proceed faster at times. To speed up chemical processes, enzymes need to have a three-dimensional structure and a chemical composition that allows them to do so. Our ability to diagnose and cure disease has been greatly aided by breakthroughs in biochemistry, and this should not come as a surprise. A variety of hereditary metabolic ailments have been connected to genetic disorders. Several disorders can be traced back to abnormal metabolic processes.

Discovering novel pharmaceutical compounds and figuring out how they operate and break down are all critical parts of biochemistry that must be explored since medications alleviate symptoms. Antibiotics like sulfonamides, penicillins, and tetracyclines are commonly used to treat bacterial infections, however, research shows that acyclovir is an excellent therapy for herpes virus infections. A lot of focus is being placed right now on the numerous intricacies of cancer carcinogenesis and cancer treatment. Nucleic acids and proteins in the body are frequently disrupted by compounds known as carcinogens, which are agents that cause cancer.

Researchers have made great strides in improving the identification of cancer-causing chemicals. Cancer prevention and treatment will surely advance as the molecular basis of the illness is more understood.

For biotechnology, molecular biology and biotechnology comprise a vital aspect of this ever-evolving science, which in turn, is a component of molecular biology (Raies &Bajic, 2016). Biochemical research has led to the development of fast, precise methods for identifying DNA and protein structural compositions. In addition, effective laboratory methods for producing genes are now being developed. If all else fails, a gene-replacement therapy may be used to cure genetic disorders. Gene therapy would be used to accomplish this.

1.4: PHYSICAL CHEMISTRY

Several disciplines of chemistry focus on materials that have similar chemical and structural properties. While some subfields of chemistry focus on a specific set of molecules, others focus on how diverse compounds interact and change over time. Since physical chemistry has been around for quite some time now, it has become an important area of study. The oldest branch of chemistry is that of physical chemistry. At a given temperature and pressure, the volume of a given gas drops proportionately. For example, the Anglo-Irish chemist Robert Boyle found this in the seventeenth century. Consequently, the product of volume V and pressure P equals a constant value—i.e., PV = constant—for gas at a constant temperature (Pritchard et al., 2019). This simple mathematical link holds for virtually all gases at ambient temperatures and pressures of one noble or less.

Thus, despite the possibility of developing more complex formulations that more accurately represent, the experimental data and the connection was found to be invalid at higher pressures. The study of chemical regularities, which are also referred to as "natural laws," lies at the heart of physical chemistry. Mathematical regularity in chemical systems was thought to be due to a web of forces and fields that existed around atoms constituting chemical elements and compounds during most of the 18th century. A quantum mechanical model of both atomic and molecular structures best explains the behavior of chemicals, according to twenty-first-century research. Theoretical chemistry, a subfield of physical chemistry, focuses on this issue in particular.

Theoretical chemists frequently use computers to aid in the resolution of complex mathematical problems. Physical chemistry also encompasses the fields of chemical thermodynamics and chemical kinetics. To better understand the relationship between chemical energy and temperature, chemical thermodynamics and chemical kinetics are employed. Electronic chemistry is a field that focuses on the interaction between electrical current and chemical change (Pinzi& Rastelli, 2019). Reversible changes in the solution's constituent elements occur when an electric current runs through it. This means that under other circumstances, the altered molecules will generate an electric current. By making an electrical connection between the battery's chemical elements, a constant voltage is generated in the battery until the chemical elements themselves run out of energy. Technology that can harness the sun's rays to fuel chemical processes capable of storing energy is attracting a lot of attention right now. The manufacturing of these gadgets might make the widespread use of solar energy more practical.

CHAPTER 2

ATOMS AND ELEMENTS

CONTENTS

2.1 INTRODUCTION TO ATOMS

Atoms are the smallest particles of matter that have all their electrons, protons, and neutrons. The more protons in an atom, the more positively charged it is and the fewer electrons it has. Atoms are made up of subatomic particles including electrons, protons, and neutrons. In an atom's nucleus, which is made up of protons and neutrons, electrons move around ceaselessly. Each atom is composed of a nucleus and one or more electrons bound to the nucleus. The nucleus is made of one or more protons and several neutrons. As it were the foremost common assortment of hydrogen has no neutrons. More than 99.94% of an atom's mass is within the nucleus. The protons have a positive electric charge, the electrons have a negative electric charge, and the neutrons have no electric charge. On the off chance that the number of protons and electrons break even, at that point the molecule is electrically neutral (Patil et al., 2008). In case an atom has more or fewer electrons than protons, at that point, it has a generally negative or positive charge, separately – such atoms are called ions. The electrons of a particle are pulled into the protons in the nuclear core by the electromagnetic force.

Figure 2.1: An image showing test for the existence of atoms.

Source:https://images.pexels.com/photos/7722830/pexels-photo-7722830.jpeg ?auto=compress&cs=tinysrgb&w=300

The protons and neutrons within the nucleus are pulled into each other by the atomic bond. This force is ordinarily more grounded than the electromagnetic force that repulses the positively charged protons from one another. Under certain circumstances, the repulsing electromagnetic force gets to be more grounded than the atomic bond. In this case, the core parts take off behind distinctive elements. This is often a basis of nuclear rot. The number of protons within the core is the nuclear number and it characterizes in which chemical component the particle has a place (Ogino et al., 2013). Particles can attach to one or more other particles by chemical bonds to make chemical compounds such as particles or gems. The capacity of atoms to connect and detach is dependent on most of the physical changes observed in nature. Chemistry is the principle that studies these changes.

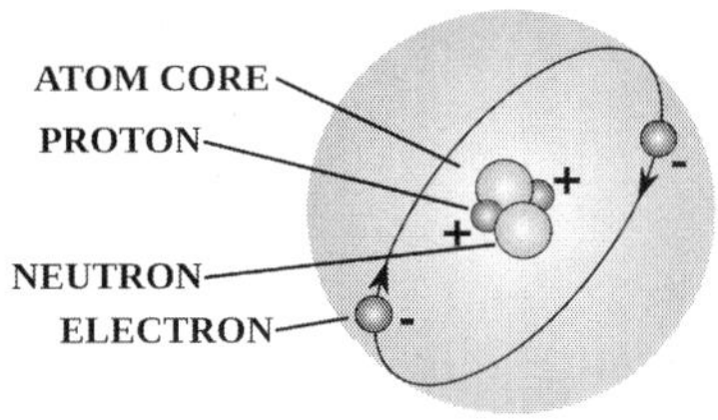

Figure 2.2: An image showing the three elements that make up an atom.

Source: https://en.m.wikipedia.org/wiki/File:Atom-struc.svg

The first atoms were discovered by ancient Greeks, who noticed that when they rubbed two pieces of amber together, they created static electricity. This led them to believe that things like air and sunlight were made up of tiny particles called "atoms." The word "atom" comes from Greek, meaning "indivisible." Within the early 1800s, the English chemist John Dalton compiled exploratory information accumulated by himself and other researchers and found a design presently known as the "law of numerous extents". He took note that in chemical compounds which contain a specific chemical component, the substance of that component in these compounds will contrast by proportions of entire little numbers. This design proposed to Dalton that each chemical component combines with other elements by a few c constant units of mass, and he chose to call these units "atoms".

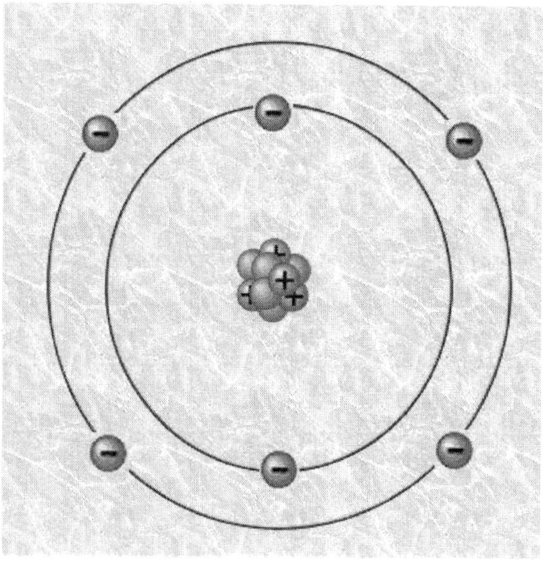

Figure 2.3: An image showing the atomic structure of a carbon atom.

Source: https://commons.wikimedia.org/wiki/File:Carbon-atom.jpg

2.2 INTRODUCTION TO ELEMENTS

Elements are substances that cannot be broken down by chemical means into simpler substances. Elements can be combined with other elements to form compounds, but they cannot be broken down into simpler substances by chemical means. Elements are usually represented by a symbol or name with a subscript number indicating how many atoms of each element make up one molecule of the compound. For example, all hydrogen atoms contain one proton in their nucleus and all carbon atoms contain six protons in their nucleus; therefore, all hydrogen atoms are isotopes of each other but not identical to any other kind of hydrogen atom or carbon atom; because they do not have the same number of neutrons.

The atomic number of a component is broken even with the number of protons in each atom and characterizes the element. For case, all carbon atoms contain 6 protons in their nuclear core; so the atomic number of carbon is 6.

Carbon atoms may have distinctive numbers of neutrons; atoms of the same component having distinctive numbers of neutrons are known as isotopes of the element (O'Gorman et al., 2012). The number of protons within the nuclear core decides its electric charge, which in turn decides the number of electrons of the molecule in its non-ionized state. The electrons are put into nuclear orbitals that decide the atom's different chemical properties. The chemical characteristics of an element are often only very slightly affected by the amount of neutrons in a nucleus.

Isotopes are atoms of the same component, that's, with the same number of protons in their nuclear core, but having diverse numbers of neutrons. Most naturally occuring elements have more than one constant isotope. The isotopes of a given component are often chemically indistinguishable, except for hydrogen, whose isotopes differ greatly from one another in relative mass—enough to have an impact on chemical reactions. All of the elements have a few radioactive isotopes, radioisotopes, even though not all of these radioisotopes happen normally. The radioisotopes ordinarily decay into other elements upon transmitting an alpha or beta molecule. In case a component has isotopes that are not radioactive, these are named "constant" isotopes. All of the known constant isotopes happen normally. The numerous radioisotopes that are not found in nature have been characterized after being misleadingly made.

The history of the disclosure and utilization of the elements started with primitive human social orders that found local minerals like carbon, sulfur, copper, and gold; in spite of the fact that the concept of a chemical component was not however caught on. Endeavors to classify materials such as these come about within the concepts of classical elements, speculative chemistry, and different comparative hypotheses all through human history. Much of the present-day understanding of elements was created from the work of Dmitri Mendeleev, a Russian chemist who distributed the primary recognizable periodic table in 1869. This table organizes the elements by expanding atomic numbers into columns in which the columns share repeating physical and chemical properties (Aramwit et al., 2010). The occasional table summarizes different properties of the elements, allowing chemists to infer bonds between them and to create expectations around compounds and potential modern ones.

Reiben	Gruppo I. — R'0	Gruppo II. — R0	Gruppo III. — R'0'	Gruppo IV. RH' R0'	Gruppo V. RH' R'0'	Gruppo VI. RH' R0'	Gruppo VII. RH R'0'	Gruppo VIII. — R0'
1	II = 1							
2	Li = 7	Bo = 9,4	B = 11	C = 12	N = 14	O = 16	F = 19	
3	Na = 23	Mg = 24	Al = 27,3	Bi = 28	P = 31	S = 32	Cl = 35,5	
4	K = 39	Ca = 40	- = 44	Ti = 48	V = 51	Cr = 52	Mn = 55	Fo = 56, Co = 59 Ni = 59, Cu = 69
5	(Cu = 63)	Zn = 65	- = 68	- = 72	As = 75	So = 73	Br = 80	
6	Rb = 86	Sr = 87	?Yt = 88	Zr = 90	Nb = 94	Mo = 96	- = 100	Ru = 104, Rh = 104, Pd = 106, Ag = 108
7	(Ag = 108)	Cd = 112	In = 113	Su = 188	Sb = 122	To = 125	J = 127	
8	Ca = 133	Ba = 137	?Di = 138	?Co = 140	—	—	—	— — — —
9	(—)	—	—	—	—	—	—	
10	—	—	?Er = 178	?La = 180	Ta = 182	W = 184	—	Os = 195, Ir = 197, Pt = 198, Au = 199
11	(Au = 199)	Hg = 200	Tl = 204	Pb = 207	Bi = 208	—	—	
12	—	—	—	Th = 231	—	U = 240	—	— — — —

Figure 2.4: Mendeleev's periodic table.

Source: By The original uploader was Den fjättrade ankan at Swedish Wikipedia. - Källa:Dmitrij Ivanovitj Mendelejev (1834 - 1907). Originally from sv.wikipedia; description page is/was here. Public Domain, https://commons.wikimedia.org/w/index.php?curid=1574857

2.3 RELATIONSHIP BETWEEN ATOMS AND ELEMENTS

Whereas numerous materials in nature do come in their pure natural form, there are other substances that do not. Instead, they are combinations of elements. When elements combine, they shape molecules. A molecule could be a group of two or more particles chemically bonded together. Molecules may be basic, or they may be very complex. For case, a few basic atoms incorporate the oxygen you breathe, which is made of two oxygen atoms, the salt you put on your food, which is made of one sodium atom and one chlorine molecule, or the water you drink, which is made of two hydrogen atoms and one oxygen atom (Neese, 2012). A more complex atom would be something like glucose, which may be a lump of sugar that has various carbon, hydrogen, and oxygen particles bonded together. Molecules are groups of two or more atoms, but there's an extraordinary sub-group of atoms, as well.

These are compounds, and a compound may be a group of two or more diverse molecules chemically bonded together. The key here is that compounds are made of at slightest two distinctive elements. This would therefore include your water because it contains both hydrogen and oxygen as well as your salt because it normally contains both sodium and chlorine.

However, the oxygen you breathe isn't a compound because it contains as if it were oxygen. But it is still an atom since it's a combination of at slightest two atoms. This could be a bit befuddling since all compounds are atoms, but not all particles are compounds.

2.4 STRUCTURE OF AN ATOM

2.4.1 Subatomic Particles

Even though the word atom initially signified a molecule that cannot be divided into smaller particles, in present-day scientific use the atom is composed of different subatomic particles. The constituent particles of an atomare the electron, the proton, and the neutron. The electron is by the distance the slightest of these particles with $9.11\times10-31$ kg, with a negative electrical charge and an estimate that's as well little to be measured utilizing accessible techniques. It was the lightest molecule with a positive rest mass measured, until the disclosure of neutrino mass (Aidas et al., 2014). Under standard conditions, electrons are bound to the positively charged nucleus by the attraction made from inverse electric charges. When an atom has more or fewer electrons than its nuclear number, it gets to be separately oppositely or positively charged as an entirety; a charged particle is called an ion.

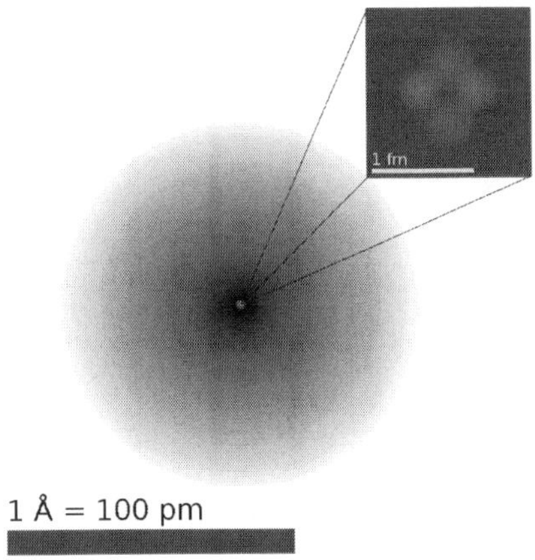

Figure 2.5: An image showing the structure of an atom: A depiction of the atomic structure of the helium atom. The darkness of the electron cloud corresponds to the line-of-sight integral over the probability function of the 1s atomic orbital of the electron. The magnified nucleus is schematic, showing protons in pink and neutrons in purple. In reality, the nucleus (and the wavefunction of each of the nucleons) is also spherically symmetric and 1s, and the four particles, each

with a different quantum number, like the electrons in the helium atom, are all most likely to be found in the same space, at the exact center of the nucleus. (For more Complex nuclei this is not the case.Thanks to Åke Back.).

Source: By User:Yzmo - Own work, CC BY-SA 3.0, https://commons.wikimedia.org/w/index.php?curid=2246091

Protons have a positive charge and a mass of 1,836 times that of the electron, at 1.6726×10−27 kg. The number of protons in a molecule is called its nuclear number. Ernest Rutherford observed that nitrogen beneath alpha-particle assault launches what showed up to be hydrogen nuclei. By 1920 he had acknowledged that the hydrogen core may be a particular molecule inside the particle and named it proton. Neutrons have no electrical charge and have a free mass of 1,839 times the mass of the electron, or 1.6749×10−27 kg. Neutrons are the heaviest of the three constituent particles, but their mass can be decreased by the charge. Neutrons and protons, which are collectively known as nucleons, have comparable dimensions, on the order of 2.5×10−15 m, although the 'surface' of these particles isn't strongly characterized (Nwodo et al., 2012).

Within the Standard Demonstration of material science, electrons are really basic particles with no internal structure, while protons and neutrons are composite particles composed of basic particles called quarks. There are two sorts of quarks in atoms, each having a fragmentary electric charge. Protons are composed of two up quarks and one down quark. Neutrons comprise one up quark and two down quarks. This refinement accounts for the distinction in mass and charge between the two particles. The quarks are held together by the solid interaction, or solid bond, which is interceded by gluons. The protons and neutrons, in turn, are held to each other within the nucleus by the atomic bond, which may be a residuum of the solid bond that has somewhat different range-properties. The gluon could be a part of the family of gauge bosons, which are basic particles that intervene in physical powers.

2.4.2 The Nucleus of an Atom

All the bound protons and neutrons in a particle make up a small atomic centre and are collectively called nucleons. The span of a centre is around the rise to femtometres, where A is the overall number of nucleons. This will be much smaller than the span of the particle. The nucleons are bound together by a short-ranged engaging potential called the remaining strong force. At partitions, a more diminutive force is much more compelling than the electrostatic force that causes decidedly charged protons to spurn each other (Moseti et al., 2016). Particles of the same component have the same number of protons, called the atomic number. Interior a single component, the number of neutrons may alter, deciding the isotope of that component. The complete number of protons and neutrons choose the nuclide. The number of neutrons relative to the protons chooses the robustness of the centre, with some specific isotopes undergoing radioactive decay.

The proton, the electron, and the neutron are classified as fermions. Fermions comply with the Pauli Exclusion Principle which disallows indistinguishable fermions, such as different protons, from involving the same quantum state at the same time. In this way, each proton within the core must possess a quantum state diverse from all other protons, and the same applies to all neutrons of the core and all electrons of the electron cloud. A nucleus with a diverse number of protons than neutrons can drop to a lower charge state through a radioactive decay that causes the number of protons and neutrons to bond more closely. As a result, molecules with coordinating numbers of protons and neutrons are stable against decay, but with expanding nuclear numbers, the mutual repulsion of the protons requires an expanding extent of neutrons to preserve the soundness of the nucleus.

2.4.3 The Electron Cloud within an Atom

The electrons in a molecule are pulled into the protons within the nucleus by the electromagnetic bond. This force ties the electron's interior to an electrostatic potential well surrounding the littler nucleus, which implies that an outside source of energy is required for the electron to escape. The closer an electron is to the nucleus, the more noteworthy the attraction force. Thus, electrons bound close to the centre of the potential well require more charge to escape than those at more noteworthy separations. Electrons, like other particles, have the properties of both a particle and a wave (Arif et al., 2010). The electron cloud could be a position interior of the potential well where each electron shapes a sort of three-dimensional standing wave, a wave shape that does not move relative to the nucleus. This behaviour is characterized by a nuclear orbital, a numerical work that characterizes the likelihood that an electron shows up to be at a specific area when its position is measured.

Each atomic orbital compares to a specific ground level of the electron. The electron can change its state to the following energy level by holding a photon with satisfactory energy to boost it into the unused quantum state. Besides, through unbonded emission, an electron within another ground state can drop to a lower ground state while exuding the overabundance of ground as a photon. These characteristic ground values, characterized by the contrasts inside the energies of the quantum states, are reliable for atomic spectral lines.

2.5 PROPERTIES OF ATOMS

Any two particles with an random number of protons in their centres are said to have the same chemical composition.Particular isotopes of the same elemenyt are atoms with increasing protons but a distinct number of neutrons. For case, all hydrogen atoms have one proton, but isotopes exist with no neutrons; hydrogen-1, by far of the preeminent common form, as well called protium, one neutron, two neutrons, and more than two neutrons (Milo et al., 2010). The known elements shape a set of atomic numbers, from the single-proton component hydrogen up to the 118-proton component oganesson molecule. All known isotopes of elements with atomic numbers more unmistakable than 82 are radioactive, in show disdain toward the truth that the radioactivity of component 83 is so slight as to be insignificant.

The stability of isotopes is influenced by the proportion of protons to neutrons, conjointly by the nearness of certain "magnetic numbers" of neutrons or protons that speak of closed and filled quantum shells. These quantum shells compare to a set of charge levels inside the shell show of the nucleus; filled shells, such as the filled shell of 50 protons for tin, confers stability on the nuclide. Of the 252 known constant nuclides, as it were four have both an odd number of protons and an odd number of neutrons: hydrogen-2, lithium-6, boron-10, and nitrogen-14. Moreover, as if it were four occuring, radioactive odd-odd nuclides have a half-life over a billion a long time: potassium-40, vanadium-50, lanthanum-138, and tantalum-180m. Most odd-odd nuclei are exceedingly unconstant about beta decay, since the decay products are even-even, and are hence more unequivocally bound.

The huge larger part of an atom's mass comes from the protons and neutrons that make it up. The full number of these particles, called nucleons in a given atom is called the mass number. It may be a positive number and dimensionless, rather than having a measurement of mass, since it communicates a number. The real mass of an atom at rest is frequently expressed in daltons (Da), also called asthe bound together nuclear mass unit (u). Hydrogen-1 which is the lightest isotope of hydrogen and is additionally the nuclide with the most reduced mass has a nuclear pressure of 1.007825 Da. The esteem of this number is called the nuclear mass (Barbatti et al., 2014). A given particle has a nuclear mass that roughly rises to its mass number times the nuclear mass unit, but this number will not be precisely numbered.

Molecules need a well-defined external boundary, so their measurements are ordinarily portrayed in terms of a nuclear sweep. Typically a degree of the separate out to which the electron cloud expands from the nucleus. This accepts the atom to show a circular shape, which is as it was compiled for particles in vacuum or free space. Nuclear radii may be determined from the separations between two nuclei when the two atoms are joined in a chemical bond. The sweep changes with the area of an atom on the nuclear chart, the sort of chemical bond, the number of neighbouring particles, the coordination number, and a quantum mechanical property known as spin. On the periodic table of the elements, molecule estimate tends to extend when moving down columns but diminishes when moving over lines, cleared out to right.

When subjected to outside strengths, like electrical areas, the shape of a molecule may veer off from round symmetry. The distortion depends on the field size and the orbital sort of external shell electrons, as appeared by group-theoretical contemplations. Aspherical deviation may be evoked for occurrence in gems, where expansive crystal-electrical areas may happen at low-symmetry cross-section positionareas (Metz et al., 2014). Critical ellipsoidal occurings have appeared to happen for sulfur particles and chalcogen particles in pyrite-type compounds. Atomic measurements are thousands of times littler than the wavelengths of light so they cannot be seen utilizing an optical microscope, even though individual particles can be observed by employing a filtering tunnelling magnifying instrument.

Each component has one or more isotopes that have unconstant nuclei that are subject to radioactive decay, causing the nucleus to transmit particles or electromagnetic radiation. Radioactivity can happen when the span of a nucleus is expansive compared with the span of the solid bond, which as it were acts over separations on the order of 1 FM. The foremost common shapes of radioactive decay are Alpha beta which is a process caused when the nucleus emits an alpha particle, which may be a helium nucleus comprising of two protons and two neutrons. The result of the outflow may be a modern component with a lower nuclear number. Beta decay and electron capture forms are controlled by the weak bond and result from a change of a neutron into a proton or a proton into a neutron. The neutron to proton move goes with by the outflow of an electron and an antineutrino, whereas proton to neutron moves, but in electron capture, causes the emission of a positron and a neutrino. The electron or positron outflows are called beta particles. Beta decay either increments or diminishes the nuclear number of the nucleus by one. Electron capture is more common than positron outflow since it requires less charge. In this sort of decay, an electron is ingested by the nucleus, instead of a positron transmitted from the nucleus. A neutrino is still radiated in this handle, and a proton changes to a neutron.

Gamma decay is a process that comes about from an alteration within the charge level of the nucleus to a lower state, coming about within the outflow of electromagnetic radiation. The energized state of a nucleus that comes about in gamma outflow as a rule happens to take after the emission of an alpha or a beta molecule. In this way, gamma decay ordinarily takes after alpha or beta decay. Other more uncommon sorts of radioactive decay incorporate the launch of neutrons or protons or clusters of nucleons from a nucleus or more than one beta molecule. An analog of gamma outflow which permits energized nuclei to lose charge distinctively, is inside conversion, a process that produces high-speed electrons that are not beta rays, taken after by a generation of high-energy photons that are not gamma rays (Bacellar et al., 2015). Some huge nuclei detonate into two or more charged parts of changing masses additionally a few neutrons, in a decay.

Basic particles have an inborn quantum mechanical property known as the spin. This can closely resemble the precise force of a protest that's spining around its centre of mass, in spite of the fact that entirely talking these particles are accepted to be point-like and cannot be said to be pivoting. Spin is measured in units of the Planck's constant (ħ), with electrons, protons, and neutrons all having spined. In an atom, electrons in movement around the nucleus have orbital precise force in expansion to their spin, whereas the nucleus itself has precise force due to its atomic spin. The magnetic field created by an atom, its magnetic moment, is decided by these different shapes of precise energy, fair as a rotating charged question classically produces an magnetic field, but the foremost overwhelming commitment comes from electron spin.

Due to the nature of electrons to comply with the Pauli Exclusion Principle, in which no two electrons may be found within the same quantum state, bound electrons combine up with each other, with one part of each match in a spin-up state and the other within the inverse, spin down state. Thus these spins cancel each other out, decreasing the entire magnetic dipole moment to zero in a few atoms with indeed several electrons. In ferromagnetic elements such as press, cobalt, and nickel, an odd number of electrons leads to an unpaired electron and a net in a general magnetic moment (Bennett et al. 2010).The orbitals of neighbouring atoms cover and a lower charge state is accomplished when the spins of unpaired electrons are adjusted with each other. When the magnetic moments of ferromagnetic molecules are lined up, the material can create a quantifiable visible field. Paramagnetic materials have molecules with magnetic moments that line up in irregular bearings when no magnetic field is available, but the magnetic moments of respective atoms follow each other in a magnetic field.

The nucleus of an atom will have no spin when it has indeed numbers of both neutrons and protons, but for other cases of odd numbers, the nucleus may have a spin. Regularly nuclei with spin are adjusted in random, but for certain elements it is conceivable to polarize a critical extent of the atomic spin states so that they are adjusted within the same direction, a condition called hyperpolarization. This has imperative applications in magnetic resonance imaging.

The potential charge of an electron in a molecule is negative relative to when the removal from the nucleus goes to limits; its reliance on the electron's position comes to the least interior of the nucleus. Within the quantum-mechanical demonstration, a bound electron can possess as it were a set of states centered on the nucleus, and each state compares to a particular charge level; see time-independent Schrödinger condition for a hypothetical clarification.

An energy level can be measured by the sum of charge required to unbind the electron from the molecule and is ordinarily given in units of electronvolts (eV). The most reduced charge state of a bound electron is called the ground state, that is, the stationary state, whereas an electron moving to the next level comes about in an excited state. The electron's charge increments together within since they separate to the nucleus increments (Mendoza & Núñez, 2009). Reliance of the charge on ℓ is caused not by the electrostatic potential of the nucleus, but by the interaction between electrons. For an electron to move between two distinctive states, e.g. ground state to begin with excited state, it must retain or emanate a photon at a charge coordinating the distinction within the potential charge of those levels, concurring to the

Niels Bohr experiment, what can be calculated by the Schrödinger condition. Electrons hop between orbitals in a particle-like mold.

2.6 PROPERTIES OF AN ELEMENT

Some terms are commonly utilized to characterize the common physical and chemical properties of the chemical elements. To start with capability is between metals, which instantly conduct charggge, nonmetals, which do not, and a small group, having midway properties and as often as possible carrying on as semiconductors. A more refined classification is routinely shown up in colored presentations of the periodic table (Mansour et al., 2010). This system limits the terms "metal" and "nonmetal" because they were certain of the more broadly characterized metals and nonmetals, counting additional terms for certain sets of the more broadly seen metals and nonmetals. The classification utilized inside the periodic tables joins actinides, alkali metals, base metals, lanthanides, transition metals, post-transition metals, metalloids, nonmetals, and noble gasses.

Another commonly utilized essential qualification among the elements is their state of matter, whether strong liquid, or gas, at a chosen standard temperature and pressure (STP). Most of the elements are solids at standard temperatures and atmospheric pressure, whereas a few are gases.

Melting and boiling point, regularly expressed in degrees Celsius at a pressure of one atmosphere, are commonly utilized in characterizing the different elements. Whereas known for most elements, either or both estimations is still undetermined for a few of the radioactive elements accessible in as it were little amounts. Since helium remains a liquid indeed at outright zero at barometrical pressure, it has a boiling point, and not a melting point.

The thickness at chosen standard temperature and pressure (STP) is regularly utilized in characterizing the elements. Thickness is frequently expressed in grams per cubic centimeter. Since a few elements are gasses at standard temperatures, their densities are as a rule expressed for their vaporous shapes; when melted or cemented, the vaporous elements have densities comparative to those of the other elements (Marian, 2012). When a component has allotropes with diverse densities, one agent allotrope is regularly chosen in outline introductions, whereas densities for each allotrope can be expressed where more detail is given.

The elements considered to date as strong tests have eight sorts of crystalstructures: cubic, body-centered cubic, face-centered cubic, hexagonal, monoclinic, orthorhombic, rhombohedral, and tetragonal. For a few of the artificially created transuranic elements, accessible tests have been as well little to decide crystalstructures.

Chemical elements may too be categorized by their beginning on Soil, with the primary 94 considered actually occuring, whereas those with nuclear numbers after 94 are as the artificial items of man-made atomic reactions (Congrains et al., 2013). Both the naturally occurring elements and the artificially occurring elements can be grouped under the same category based on the above-mentioned properties.

Conclusion

Atoms and elements are related in one way or the other in the sense that; the different elements which exist are made of atoms; with the only difference being that; each element has a different atomic structure. In summary, all elements are collections of millions of atoms.

CHAPTER 3

CHEMICAL COMPOUNDS

CONTENTS

Chemical compounds (or molecular groups) are bound together by chemical bonds and composed of materials derived from diverse sources. As a result, a compound is not a molecule composed of atoms from a single element.

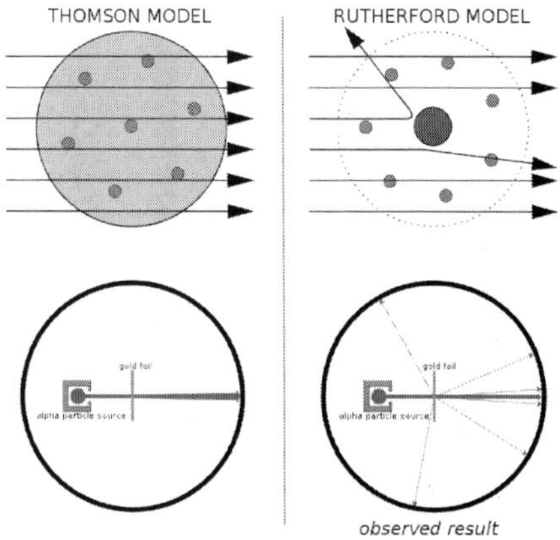

Figure 3.1: Evolution of atomic structure.

Source: By Kurzon - Own work, CC BY-SA 3.0, https://commons.wikimedia.org/w/index.php?curid=32215297

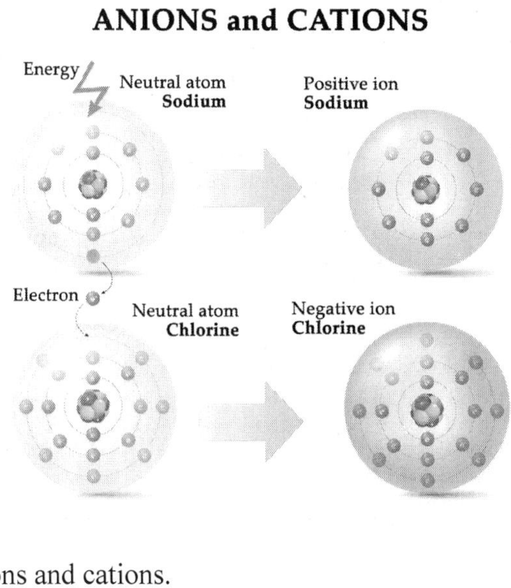

Figure 3.2: Anions and cations.

Source: https://www.istockphoto.com/vector/anions-and-cations-for-example-sodium-and-chlorine-atoms-gm611074184-105067087

There are four fundamental types of compounds based on how the atoms in a compound are connected:

- Covalent bonds keep molecular bonding together.
- Compounds are held together by ionic bonding.
- Metallic bonds hold intermetallic complexes together.
- Coordinated covalent bonds hold some complexes together.

In a chemical formula, the conventional element names and numerical subscripts define the number of atoms present in each element of a complex molecule. H_2O, for example, indicates that two hydrogen atoms are bonded to a single oxygen atom (Matranga & Corsi, 2012). Numerous chemical compounds have been given distinct CAS numbers by the Chemical Abstracts Service (CAS). Around the world, 350,000 chemical compounds have been registered for manufacture and consumption.

A chemical reaction occurs when one molecule interacts with another to produce a new chemical compound. A break in one or both two atom-to-atom bonds may occur, resulting in the creation of new bonds.

3.1 THE CLASSIFICATION OF COMPLEX COMPOUNDS

A multitude of characteristics can be used to classify chemical compounds. The existence of particular elements is one prevalent feature. Oxides, hydrides, and halides, for example, are all made up of one or more halogen atoms (Group 17). To be categorized as organic, a material must have a carbon atom backbone, whereas all other compounds are classified as inorganic. An organometallic compound is an organic compound that contains metal atoms.Ionic bonds exist in sodium chloride, also known as table salt.

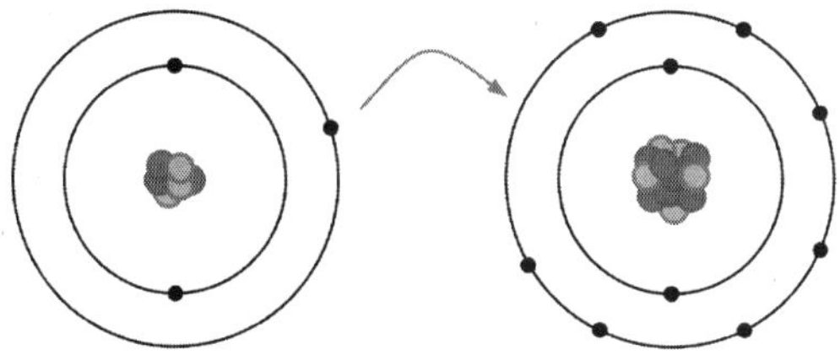

Figure 3.3: Examples of an ionic bond.

Source: By JackFromReedsburg - Own work, CC0, https://commons.wikimedia.org/w/index.php?curid=97865879

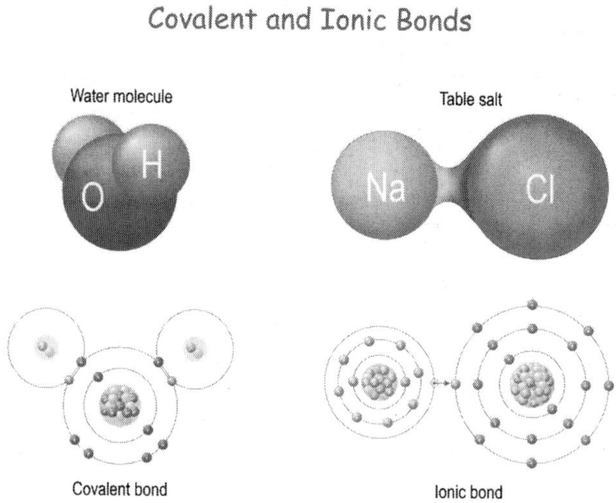

Figure 3.4: Covalent and ionic bonds.

Source: https://www.istockphoto.com/vector/ionic-vs-covalent-bonds-gm1301181932-393288502

Sodium chloride exhibits ionic bonding. During chemical processes, one electron from sodium (Na) is exchanged for one from chlorine (Cl) to generate a stable ionic molecule (sodium chloride; ordinary table salt).

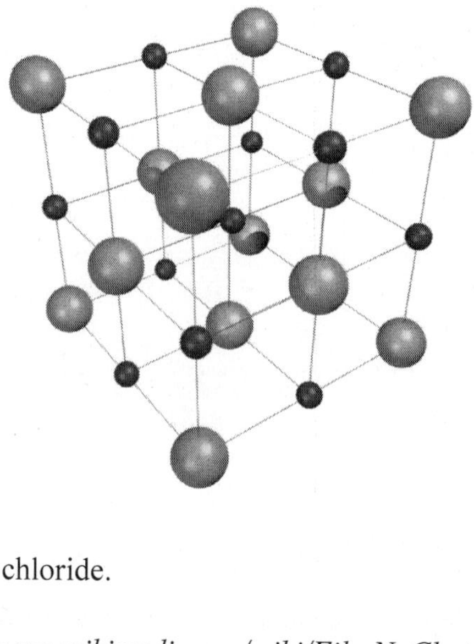

Figure 3.5: Sodium chloride.

Source: https://commons.wikimedia.org/wiki/File:NaCl-structure.jpg

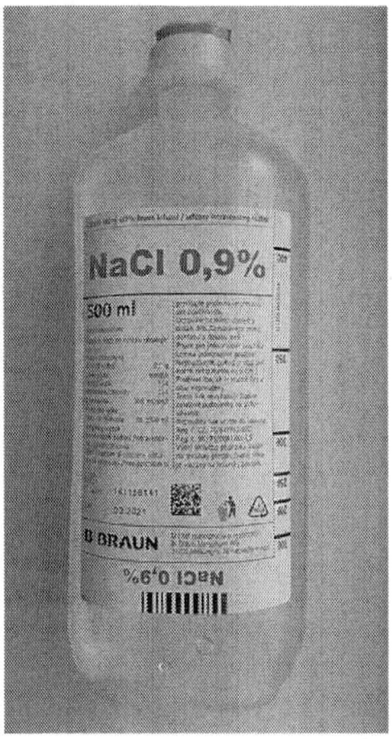

Figure 3.6: NaCl in its liquid state.

Source: By Wesalius - Own work, CC BY 4.0, https://commons.wikimedia.org/w/index.php?curid=69066763

Chemical compounds are also classified into subgroups based on the types of bonds they possess. Two ions with opposing charges attract one another in an ionic combination. Common salt is one of the most well-known ionic compounds (sodium chloride). Molecular compounds are built up of atoms that share electrons (covalent bonding) (Lovley, 2003). Water, methane, and hydrogen fluoride are made up of H_2O, CH_4, and HF molecules, respectively.

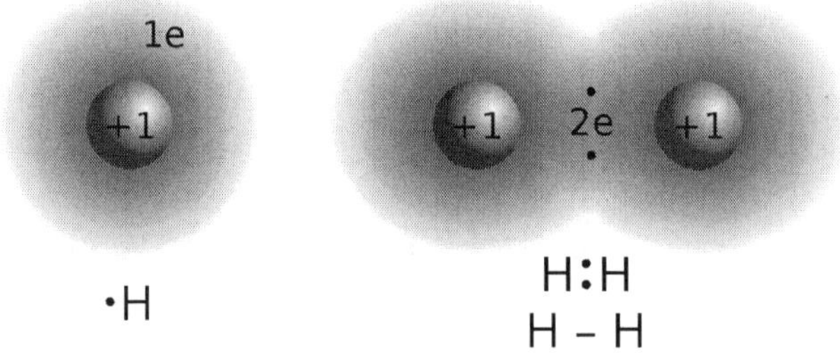

Figure 3.7: How bonding occurs covalently.

Source: By Jacek FH - Own work, CC BY-SA 3.0, https://commons.wikimedia.org/w/index.php?curid=2781099

A third approach of categorizing is based on the compounds' predicted reactivity or the sorts of chemical reactions they are expected to undergo. 'Acids, for example, are chemicals that when dissolved in water create H^+ ions (protons) and aqueous solutions. When addressing the chemical characteristics of acids, they are referred to as proton donors. The most typically encountered acids are HCl, H_2SO_4, HNO_3, and H_3PO_4 (potassium phosphate) in aqueous solutions (phosphoric acid). Proton acceptors, on the other hand, are bases. The hydroxide ion (OH), which interacts with an H^+ ion to generate a water molecule, is the most common base. $H^+ + OH^-$ (usually written H_2O)

Another major family of chemical processes is oxidation-reduction reactions. The oxidation process results in electron loss, whereas the reduction process results in electron gain. Electrons (e) are transported from sodium atoms to chlorine atoms in the reaction 2Na + Cl_2 2NaCl to create Na^+ and Cl^- ions in the reaction product sodium chloride (Campos & Vasconcelos, 2010).

Each sodium atom loses an electron and is oxidized in this process, whereas each chlorine atom acquires an electron and is reduced. The reducing agent in this reaction is sodium, which produces electrons, while the oxidizing agent is chlorine (it consumes electrons). Metals are the most prevalent reducing agents due to their tendency to lose electrons in nonmetal processes. Halogens such as fluorine (F_2), chlorine (Cl_2), and bromine (Br_2), as well as particular oxy-anions such as permanganate (MnO_4) and dichromate ($Cr_2O_7{}^{2}$-), are the most prevalent oxidizing agents

3.2 COMPOUNDS THAT DO NOT HAVE A CARBON-HYDROGEN BOND

Inorganic compounds are those that include two or more atoms other than carbon, as well as those that contain carbon but lack carbon-carbon bonds, such as cyanides and carbonates. The elements or element groups that inorganic compounds comprise are the most often used approach for categorizing them. Oxides, for example, might be ionic or molecular. Ionic oxides are made up of O_2 (oxide) ions and metal cations, whereas molecular oxides are made up of molecules with oxygen (O) covalently bonded to nonmetals like sulfur (S) or nitrogen (N) (Lischka et al., 2011). When ionic oxides dissolve in water, oxygen ions (O_2) react with water molecules to generate hydroxide ions (OH), resulting in a basic solution. When molecular oxides react with water, oxyacids such as sulfuric acid (H_2SO_4) and nitric acid (HNO_3) are formed (HNO_3). Hydrides (carrying hydrogen atoms or H ions), nitrides (containing N_3 ions), phosphides (containing P_3 ions), and sulfides (containing S_2 ions) are further examples of inorganic substances (which contain S2 ions). An example of an inorganic compound is borane.

Transition metals may be used to make a wide range of inorganic compounds. The most important are coordination compounds, which have two to six ligands encircling the metal atom or ion. Ligands are ions or neutral substances that may contribute electron pairs to metal atoms, forming a covalent coordination link.Because just one substance (the ligand) contributes both electrons shared in the bond, the resultant covalent bond is given a name. $[Co(NH_3)_6]Cl_3$ is a coordination compound composed of the $Co(NH_3)_6^{3+}$ ion, a cobalt ion (Co^{3+}), and six ammonia molecules (NH_3) linked as ligands.

There was no systematic technique for naming compounds in the early days of the study of chemistry. Chemists gave common compounds names like lead sugar, quicklime, milk of magnesia, Epsom salts (see magnesium), and laughing gas. These include names that are common or inconsequential. As chemistry progressed, it became clear that adopting common names for all known compounds, which numbered in the millions, would cause significant confusion. Remembering the insignificant names of so many compounds would be difficult. As a result, a system of nomenclature (naming) has been devised (Lisio et al., 2019). Certain well-known compounds, on the other hand, are invariably referred to by their common names. For example, the scientific designations H_2O and NH_3 are rarely used; these important substances are often referred to as water and ammonia, respectively.

The most fundamental chemical compounds are two-element compounds, often known as binary compounds. Because binary ionic compounds and binary molecular (covalent) compounds have different naming criteria, they will be addressed separately.

3.3 BINARY-STRUCTURED COMPOUNDS

3.3.1 Binary Ionic Compounds

Binary ionic compound nomenclature is as simple as identifying the ions using the following principles. The positive ion (cation) is detected first, followed by the negative ion (anion). The term the parent element is used to refer to a simple cation (consisting of a single atom). In the nomenclature of substances containing this ion, for example, Li^+ is referred to as lithium. Similarly, sodium is denoted as Na^+, magnesium as Mg^{2+}, and so on.

The suffix -ide is appended to the root of the parent element's name (obtained from a single atom) to denote a simple anion (obtained from a single atom). As a result, the F^- ion is referred to as fluoride, Br^- as bromide, S^{2-} as sulfide, and so on.

The following are some instances of binary ionic compound nomenclature rules:

Compound ions present name

NaCl Na^+, Cl^- sodium chloride

KI, K^+, I^- potassium iodide

CaS, Ca^{2+}, S^{2-} calcium sulphide

CsBr, Cs^+, Br^-, caesium bromide

MgO, $Mg^{2-,}$ $O2^-$, magnesium oxide

Simple ions in ionic compound formulations are represented by chemical symbols. for example Cl, , has the chemical symbol Cl^-, Na has the sign Na^+, and so on. The charge is always included in the data when showing individual ions. Although the potassium and bromide ions are referred to as K^+ and Br^-, respectively, the KBr formula for potassium bromide is KBr.

When a metal can produce a range of cations, specifying the precise cation's charge is required. Pb^{2+} and Pb^{4+} ions, for example, can be found in lead-containing ionic compounds (Pb). Iron (Fe), tin (Sn), and gold (Au) can also produce Fe^{2+}, Fe^{3+}, Sn^{2+}, and Au^{3+} ions (Au). In binary compounds containing these metals, the charge of the ion must be represented by a Roman numeral (Laganà et al., 2019). Iron (III) chloride is the chemical compound $FeCl_{3}$, which includes Fe^{3+}. The chemical compound $FeCl_2$ contains Iron (III) chloride (Fe^{2+}). The Roman numeral in each instance's name denotes the charge.

3.3.2 Common Simple Cations and Anions

H^+

Hydrogen

H^-

Hydride

Li^+

Lithium

F^-

Fluoride

Na^+

Sodium

Cl^-

Chloride

K^+

Potassium

Br^-

Bromide

Cs^+

Cesium

I^-

Iodide

Be^{2+}

Beryllium

O^{2-}

Oxide

Mg^{2+}

Magnesium

S^{2-}

Sulfide

Ca^{2+}

Calcium

Ba^{2+}

Barium

Al^{3+}

Aluminium

Ag^+

Silver

An alternate approach for identifying compounds containing metals that produce just two ions is occasionally observed, particularly in older literature. The name of the higher-charged ion ends in -ic, whereas the name of the lower-charged ion ends in -ous. Fe^{3+} is known as the ferric ion, whereas Fe^{2+} is known as the ferrous ion (Kleber & Johnson, 2010). The compounds $FeCl_3$ and $FeCl_2$ are thus known as ferric chloride and ferrous chloride, respectively.

Mercury (I) ions are usually found together as Hg_2^{2+} when they combine to produce multiple cations. Alternative term for the systematic name of an ion

Iron (III) ferric Fe^{3+}

Iron (II) ferrous Fe^{2+}

Cu^{2+} cupric copper (II)

Cu^{+} cuprous copper (I)

Cobaltic Co^{3+}cobalt (III)

Cobalt (II) cobaltous Co^{2+}

Tin (IV) stannic Sn^{4+}

Tin (II) stannous Sn^{2+}

Plumbic Pb^{4+} Lead(IV)

Plumbous pb^{2+} Lead(II)

Mercuric Hg^{2+} Mercury(II)

Mercury(I) mercurous Hg_2^{2+}Molecular compounds that are covalent (binary).

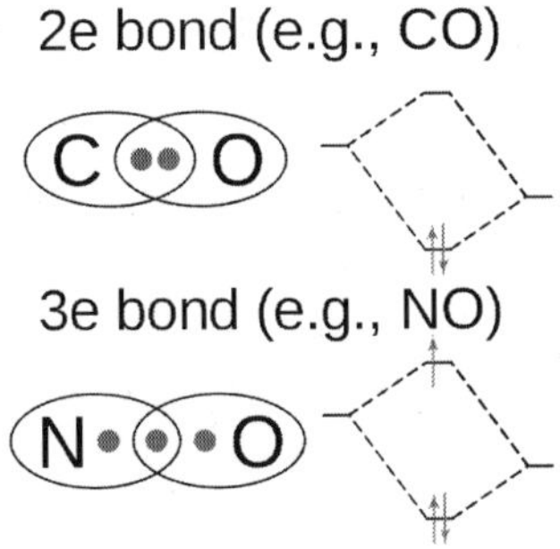

Figure 3.8: Types of covalent bonds.

Source: By Officer781 - Own work, MO diagrams from File:MO-covalent and polar bonds.svg, CC BY-SA 3.0, https://commons.wikimedia.org/w/index.php?curid=22109708

Binary molecular (covalent) compounds are produced as a result of a reaction between two nonmetals. Despite the lack of ions, these compounds are termed similarly to binary ionic compounds (Colussi et al., 2013). The rules for binary covalent compound naming are as follows:

1. The first element of the formula is presented first, followed by the element's entire name.
2. The second ingredient is called an anion.
3. Prefixes are used to denote the number of atoms present. If the first element is a single atom, the prefix mono- is omitted. CO, for example, is known as carbon monoxide rather than mono carbon monoxide.

The following examples show how the principles are applied to nitrogen and oxygen covalent compounds:

3.4 COMPOUND NAME SYSTEMATIC NAME

Nitrous oxide (N_2O) (Nitrous oxide) (laughing gas) Dinitrogen monoxide

Nitrogen monoxide (NO) is a form of nitric oxide (NO).

Nitrogen dioxide is abbreviated as NO_2.

N_2O_3 (dinitrogen trioxide) (N_2O_3)

N_2O_4 dinitrogen tetroxide (N_2O_4)

N_2O_5 (dinitrogen pentoxide) (N_2O_5)

To avoid awkward pronunciations, the final o or an of the prefix is frequently dropped when the element name begins with a vowel. N_2O_4 is called dinitrogen tetroxide, and CO is called carbon monoxide.

3.5 PREFIXES THAT APPEAR IN CHEMICAL CLASSIFICATION

As a prefix, the number of atoms

Mono- 1

Di- 2

Tri- 3

Tetra- 4

Penta- 5

Hexa- 6

Hepta- 7

Octa- 8

Non-binary compounds

Ionic compounds contain polyatomic ions. Ammonium nitrate (NH_4NO_3) is an example of an ionic compound because it contains two polyatomic ions, NH_4^+ and NO_3. As the name implies, a polyatomic ion is a charged entity composed of several atoms bound together (Krylov & Gill, 2013). Polyatomic ions have distinct names that are used to identify compounds that contain them.

Frequently encountered polyatomic ions

Bisulfate and bicarbonate are common names for hydrogen sulfate.

Respectively, hydrogen carbonate.

Ion name

Carbonate of NH_4^+ ammonium CO_3^{2-}

NO_2 nitrite HCO_3 hydrogen carbonate

ClO hypochlorite NO_3 nitrate

ClO_2 chlorite SO_3^{2-}sulfite

ClO_3 chlorate SO_4^{2-}sulfate

HSO_4 hydrogen sulfate ClO_4 perchlorate

$C_2H_3O_2$ acetate OH hydroxide

MnO_4 permanganate CN cyanide

Phosphate PO_4^{3-} $Cr_2O_7^{2-}$ dichromate

HPO_4^{2-} hydrogen phosphate CrO_4^{2-} chromate

H_2PO_4 peroxide dihydrogen phosphate O_2^{2-}

Polyatomic anions are classified into several types, each of which contains an atom of a specific element combined with varying numbers of oxygen atoms. An anion is an oxygen anion. When there are only two members in a series, the ion with the fewest oxygen atoms is called -ite, and the other ion is called -ate. SO_3^{2-} is known as sulphite, whereas SO_4^{2-} is known as sulphate. When there are more than two oxy anions in the series, the prefixes hypo- (less than) and per- (more than) are used to name the members of the series with the smallest and greatest number of oxygen atoms, respectively (Kharat et al., 2017). A good example is oxy anions containing chlorine:

Hypochlorite of ClO^-

Chlorite ClO_2

Chlorate of ClO_3

Perchlorate of ClO_4

The naming of polyatomic ionic compounds is similar to the naming of binary ionic compounds. The compound NaOH is known as sodium hydroxide because it contains the Na^+ (sodium) cation and the OH^- (hydroxide) anion. When a metal that can form multiple cations is present, the charge on the cation must be specified using a Roman numeral, just as it is in binary ionic compounds. The compound FeSO4 is known as Iron (II) sulfate because it contains Fe^{2+}.

3.6 ACIDS

An acid is a molecule that contains at least one hydrogen cation (H^+) attached to an anion. The presence or absence of oxygen in the anion determines the classification of acids. If the anion does not contain oxygen, the acid is named with the prefix hydro- and the suffix -ic. For example, hydrochloric acid is HCl dissolved in water. Hydrocyanic and Hydrosulfuric acids acids dissolved in water, respectively.

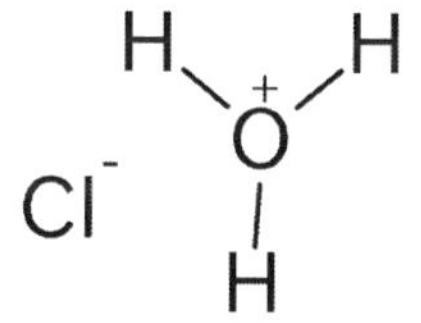

Figure 3.9: HCL acid.

Source: By LeticiaLL - Own work, CC BY-SA 4.0, https://commons.wikimedia.org/w/index.php?curid=99863027

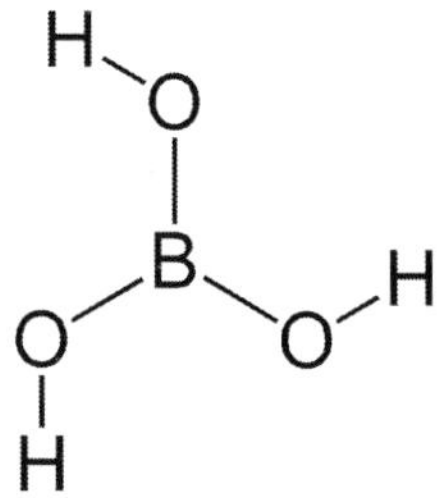

Figure 3.10: Boric acid.

Source: By Ben Mills - Own work, Public Domain, https://commons.wikimedia.org/w/index.php?curid=2664744

If the anion of the acid contains oxygen, the name is formed by appending the suffix -ic or -ous to the root name of the anion. If the anion's name ends in -ate, -ic (or sometimes -ric) (or sometimes -ric) is used instead (or sometimes -ric). H_2SO_4 is sulfuric acid due to the presence of the sulfate anion (SO_4^{2-}); H_3PO_4 is phosphoric acid due to the presence of the phosphate anion (PO_4^{3-}), and $HC_2H_3O_2$ is acetic acid due to the presence of the acetate ion ($C_2H_3O_2$) ($C_2H_3O_2$) ($C_2H_3O_2$). When designating acids with a -ite ending, the -ite is substituted with -ous. H_2SO_3, containing sulfite (SO_3^{2-}), is known as sulfurous acid, whereas HNO_2, containing nitrite (NO_2), is known as nitrous acid (Kawser Hossain et al., 2016). The acids of the oxy anions of chlorine are used to demonstrate the rules for naming acids with oxygen-containing cations.

Acids with unusual names

A type of orthoboric acid is H_3BO_3 orthoboric acid*.

Carbonate of soda (H_2CO_3)

H3PO4 orthophosphoric acid

(H4P2O7) pyrophosphoric acid (H4P2O7)

H5P3O10 triphosphoric acid

HPO3n (Metaphosphoric acid)

HPO3 trimetaphosphoric acid

Acid phosphoric (H3PO3)

(H_3PO_2) Hypophosphorous acid

(H_2SO_5) Peroxosulfuric acid

The acid with ionic $H_2S_2O_6$

Thiosulfuric acid (TSA) ($H_2S_2O_3$)

($HMnO_4$) Permanganic acid ($HMnO_4$) ($HMnO_4$)

3.6.1 Commonly Used Acids Formula Name

HF is an abbreviation for hydrofluoric acid.

A type of hydrochloric acid is HCl (hydrochloric acid).

(HBr) Hydrobromic acid

A type of hydroiodic acid is hydroiodic acid (HI).

HCN (hydrocyanic acid) (HCN)

Hydrogen peroxide (H_2S)

Nitric acid (HNO_3)

Nitrous acid (HNO_2)

Sulfuric acid H_2SO_4

Sulfuric acid H_2SO_3

$HC_2H_3O_2$ acetic acid

Anion formula acid name

Perchlorate perchloric acid $HClO_4$

Chloric acid $HClO_3$ chlorate

Chlorite chlorous acid $HClO_2$

Hypochlorous acid HClO hypochlorite

Complex ion-containing compounds

A coordination compound is composed of one or more complex structural units, each of which contains a nucleus atom that is directly bonded to a ring of groups known as ligands. These structural links serve as the foundation for the coordination compound nomenclature.

3.7 ORGANIC COMPOUNDS

Carbon (C) atoms serve as the structural underpinning in organic molecules, allowing for a vast range of possibilities. Organic molecules are essential to all living things on Earth (and most likely elsewhere in the cosmos). Organic compounds include consumables such as lipids, proteins, and carbohydrates, as well as necessities such as hemoglobin, chlorophyll, enzymes, hormones, and vitamins (Karsenti, 2008). Organic substances are also the foundation of life. A wide range of organic chemicals can be found in everyday life. Natural and synthetic fibres such as wool and silk, as well as man-made materials like nylon and polyester; wood and coal, as well as petroleum and natural gas; and the many elements of protective finishes such as paint, varnish, and lacquer are examples of these.

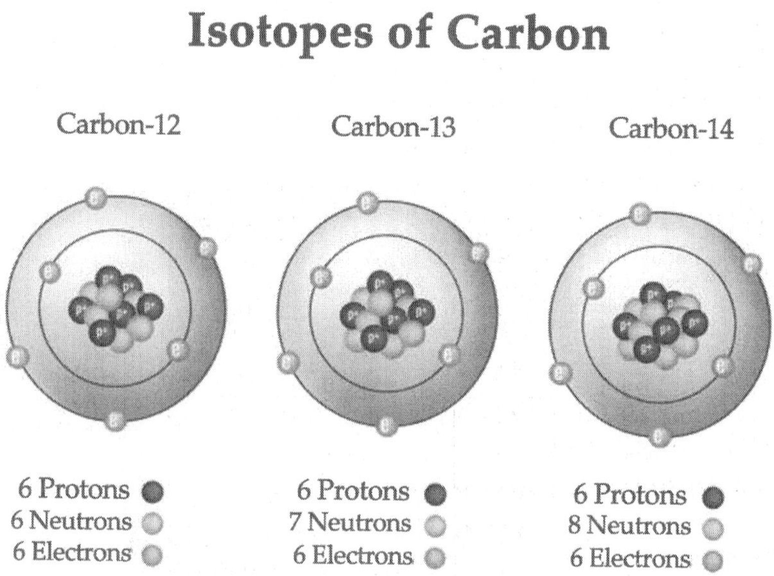

Figure 3.11: Isotopes of carbon.

Source: https://www.istockphoto.com/vector/isotopes-of-carbon-3d-vector-illustration-gm1360655261-433592347

Organic compounds can be represented using condensed, extended, or three-dimensional structural formulas.

3.8 CARBON'S FUNCTION IN CHEMISTRY

Carbon is a scarce element because it can form massive covalent bonds with other atoms and with itself. Because carbohydrates are in the second horizontal row of the periodic table, they are more likely to share electrons than to gain or lose them. Carbon has the most outer shell electrons (four) among the second-row elements for forming covalent bonds. In contrast to other elements, such as phosphorus [P] and cobalt [Co], which can only create five and six covalent bonds with other elements, respectively, carbon may form an infinite number of these bonds.) When completely bonded to other atoms, the carbon atoms four bonds form an angle of approximately 109.5 degrees with one another. Because carbon atoms may unite indefinitely to produce exceptionally heavy combinations, the resultant' molecular pressure of the molecule can take on an infinite variety of three-dimensional shapes (Chen et al., 2015). The incorporation of other elements, such as hydrogen, oxygen, nitrogen, fluorine, chlorine, bromine, and iodine, as well as sulfur (S), increases the potential for variety in organic molecules. Organic molecules are essential for the survival of life on Earth due to their extraordinary capacity to change their chemical characteristics.

Organic molecule structures are commonly expressed using simplified structural equations. These equations depict covalent bonds between atoms as well as the types and amounts of atoms in the molecule. Simple molecular equations, which just define the number and type of atoms in a molecule, do not convey this information. Because there are few atoms and just one possible arrangement of the atoms, structural equations are unnecessary for the vast majority of inorganic substances. Short lines represent covalent bonds in organic structural formulas. Chemical symbols, like molecular formulas, are used to represent atoms of certain elements.

The type of three-dimensional information conveyed by a molecule's structural formula varies widely, as does the amount of information the formula is intended to communicate. Hydrocarbons, which are among the most basic organic substances, are an excellent example of the range of complexity. The chemical formula for ethane, for example, is C_2H_6. The most basic structural formula for ethane indicates that it is made up of two carbon atoms connected by three hydrogen atoms apiece. There is no information on the three-dimensional structure of this two-dimensional picture that adequately reflects the ethane bonding pattern. A more sophisticated structural formula can be employed to more exactly portray the molecule's three-dimensional structure. This structural formula precisely depicts the tetrahedral orientation of the four atoms (one carbon and three hydrogens) connected to each carbon, as well as the molecule's distinctive design.

As carbon atoms are added to organic molecules, their size increases. Butane, for example, has the formula C_4H_{10} because it comprises a four-carbon chain and an equal amount of hydrogen atoms (10 in total). Adding more carbon atoms to a chemical framework may result in the formation of branching or cyclic patterns in the carbon chain. In the hydrocarbon cyclohexane, C_6H_{12}, six carbon atoms are organized in a tetrahedral configuration (Kalra et al., 2016). No ring structures are illustrating the hydrogen atoms that suit the bonding demands of the carbon atoms, which are generally shown as regular polygons with each vertex representing a carbon-atom atom. However, the cyclic structural polygon convention, which depicts a molecule's bonding arrangement, does not give useful information about the molecule's true 3-D structure. The polygon only contains a two-dimensional depiction of the three-dimensional molecule.

If the bonding conditions are exactly right, neighbouring atoms will form numerous bonds. When two atoms exchange two electron pairs to make two covalent bonds, a double bond is established; a triple bond is produced when three electron pairs are shared to form three covalent bonds. Many bonds have distinct structural and electrical properties in addition to their numerous chemical properties. Covalent bonds, such as ethene and C_2H_4, are made up of six atoms, with the electrons of the second covalent bond placed above and below the plane in which the atoms are arranged. Emission from the second and third covalent bonds is distributed throughout four regions on either side of the bond axis, with electrons from all three bonds aligned in a straight line.

3.9 FUNCTIONAL GROUPS

Scientists learned early on that some atomic and covalent groupings, termed "functional groups" showed distinct reactivity patterns while investigating organic molecules. As a result, even though there are over a million known organic compounds, all molecules with the same functional group have equivalent reactivity at the functional group position. As a guiding principle, functional groups are intensively used in organic chemistry (Chaplin, 2006). A molecule's functional groups can be utilized to predict and comprehend a variety of chemical processes that the molecule will go through (most compounds include more than one functional group).

Because covalent bonds between carbon and hydrogen are extremely strong and uniformly dispersed, hydrocarbons with only single carbon-hydrogen bonds are not especially volatile. One or more weak bonds or bonds with uneven electron distributions between the two atoms may increase the reactivity of the molecule. When the electrons of a connected atom are pulled closer together with the electrons of the covalent bonds two other electrons, a partially negative charge is transferred to it. In polar bonds, electron pairs are dispersed unequally among the atoms. The existence of polar and other types of electrically reactive bonds increases a molecule's chemical reaction potential. Certain molecular bonds must be dissociated for chemical reactions to occur. Attracts atoms or groups of atoms with low electron concentrations to a high electron density region. This is the initial stage in the bond-breaking process of a chemical reaction. For these reasons, atomic variations in electron density are required for chemical reactions.

Two fundamental bonding properties form reactive sites in functional groups. The first is that there are several bonds. Outside the axis of the atom-to-atom connection, both double and triple bonds have zones of high electron density. Functional groups are highly reactive atoms or groups of atoms in a molecule. Functional groups include double and triple bonds. The second type of reactive site is formed when an element other than carbon or hydrogen (known as a heteroatom) is bonded to carbon. In comparison to carbon, all heteroatoms have a higher or weaker affinity for electrons (Chemat et al., 2012). As a result, the degree of polarity of each connection between a carbon and a heteroatom is determined by the difference in electron-attracting properties of the two atoms. Significant atomic groupings with reactive polar bonds can generate functional groups.

To emphasize the universality of interactions across molecules with the same functional group, l chemists frequently use the letter R to indicate the less reactive parts of compounds. It makes no difference how intricate a molecule contains a double bond since it has the universal alkene formula.

At this position in the molecule's structure, double bond reactions will occur. The functional group reaction does not affect the molecule's other four R groups.

Polyfunctional molecules can have more Complex structures than monofunctional molecules due to the diversity and interconnectedness of their functional groups. Multiple natural compounds have a Complex three-dimensional structure with many functional groups precisely where they should be.

3.9.1 Alkanes

Alkanes are the only carbon and hydrogen molecules that are hydrocarbons (carbon and hydrogen atoms bonded together by a single bond). A shared electron pair covers the space between the two atoms in these single bonds, which is known as a sigma (σ) bond. For example, Sigma bonds may exist between two or more of the same elements, such as two or more of the same elements. The vast majority of organic molecules have an inert structure composed of alkane sequences. Alkanes are thus not classified as a functional class. An alkyl group is formed when a hydrocarbon chain is linked as a substituent to a more basic structural unit (Dubernet et al., 2010). Methane (CH_4; the principal component of natural gas), ethane (C_2H_6), propane (C_3H_8), and butane (C_4H_{10}; the liquid fuel in pocket lighters) are the four fundamental alkanes (C_4H_{10}; the liquid fuel in pocket lighters). For example, the most well-known ring-form hydrocarbon chain is cyclohexane (C_6H_{12}) (C_6H_{12}).

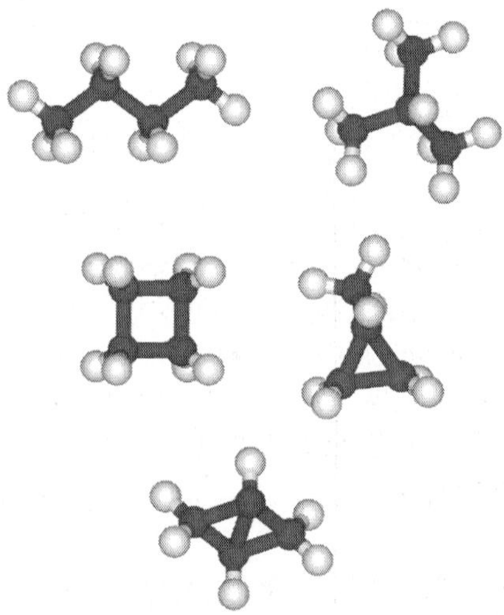

Figure 3.12: Examples of alkanes: C_4 alkanes and cycloalkanes (left to right): *n*-butane and isobutane are the two C_4H_{10} isomers; cyclobutane and methyl-cyclopropane are the two C_4H_8 alkane isomers. Bicyclo[1.1.0]butane is the only C_4H_6 alkane and has no alkane isomer; tetrahedrane (below) is the only C_4H_4 alkane and so has no alkane isomer.

Source: By Fvasconcellos - Own work, Public Domain, https://commons.wikimedia.org/w/index.php?curid=1668243

3.9.2 Alkenes

Alkenes are organic molecules with a carbon-carbon double bond. A bond is a two-atom structure that shares an electron pair. The second pair of electrons, referred to as the pi (π) bond, takes up space on both sides of the bond. This is due to the bond's electron pair being farther away from the positively charged carbon nuclei than the bond's electron pair. This results in an area of high electron density. A too strong carbon-carbon tetrahedron connection, on the other hand, will attract electron-deficient atoms, causing the link to break and new bonds to form. Adding molecular hydrogen to alkanes with just bonds is an example of an alkene reaction that demonstrates how a functional group's electrical characteristics impact its reactivity.

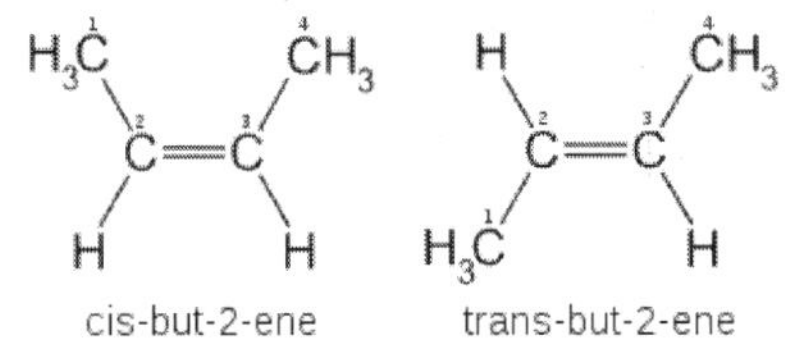

Figure 3.13: Classes of alkenes.

Source: https://commons.wikimedia.org/wiki/File:Cis-trans_example.svg

Because two carbon-hydrogen bonds are stronger than one carbon-carbon bond and one hydrogen-hydrogen bond, these processes are more energy efficient (one carbon-carbon bond and one hydrogen-hydrogen bond). An unsaturated alkene is one in which atoms are added to the alkene's bond regularly to form new bonds. The saturation of an alkane happens when it cannot be converted into a molecule with extra bonds by addition operations (Dufrêne et al., 2017).

Alkenes are an important atom in organic chemistry that may be found all over the place. Rubber, 2-methyl-1,3-butadieneisoprene (used in polyethylene manufacture), and vitamin A are just a few examples (essential for vision). In ethene, the carbon atoms of the alkene and the four atoms linked by a double bond all lie in the same plane at the same time.

3.9.3 Alkynes

Alkynes have a triple bond between their carbon atoms. When a single bond is connected by two additional bonds, a triple bond is produced. They produce electron-dense zones that run perpendicular to the carbon-carbon bond axis. Regardless of how hard we try, reactions can still disrupt carbon-carbon bonds, leading to the production of even stronger ones.

Ethyne, often known as acetylene, is the most common alkyne and is used as a welding fuel in oxyacetylene torches. Even though alkynes are scarce, the fungicide capillan has two functional groups that are alkynes.

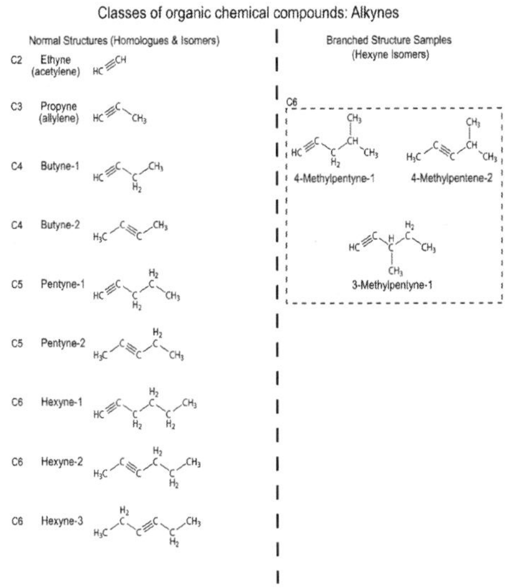

Figure 3.14: Classes of alkynes.

Source: https://www.istockphoto.com/vector/organic-chemicals-hydrocarbons-alkynes-gm483244310-70523355

3.9.4 Arenes (aromatic hydrocarbons)

Functional groups with three pairs of doubly bound atoms (often all carbon atoms) arranged in the shape of a regular planar (flat) hexagon exhibit distinct physical and chemical characteristics. A series of single and double bonds alternate with each other to form the hexagonal ring. C6H6 benzene, discovered in 1825 by English scientist and physicist Michael Faraday, may contain this functional group. Arenes include benzene (or other structurally related) rings. Aromatic hydrocarbons, such as benzene and many other bigger arenes, have long been recognized as such due to their strong odour. Because of the electrical needs of six or more pi electrons, benzene and other big arenes have a distinctive planar structure (Huang & Zou, 2010). When aromatic units are used as structural unit substitutions, they are referred to as aryl substituents. Naphthalene, the active ingredient in mothballs, is a fusion of two benzene rings. When organic molecules are burnt, a little amount of benzo[a]pyrene, an aromatic hydrocarbon, is released. This and many other polycyclic aromatic hydrocarbons are carcinogenic. Aromatic chemicals are abundant in nature. Aromatic smells, such as anisole, vanillin, and benzaldehyde, are pleasing.

CHAPTER 4

QUANTITIES OF REACTANTS AND PRODUCTS

CONTENTS

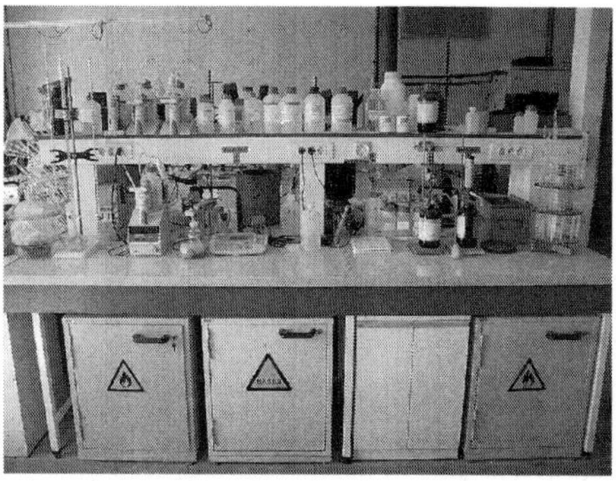

Figure 4.2: Chemical equipment in chemical analysis.

Source: By Jean-Pierre from Cosne-Cours-sur-Loire (Nièvre, Burgundy region), France - Chemistry laboratory, detail, CC BY-SA 2.0, https://commons.wikimedia.org/w/index.php?curid=17155591

As the reaction progresses, the number of products in the reaction mixture grows. Various types of reactions and their products.

Combustion Reactions: The results of combustion reactions are normally carbon dioxide and water for complete combustion (of hydrocarbons) and carbon monoxide for partial combustion.

- Decomposition Processes: Decomposition processes yield molecules that are smaller than their reactants.
- Acid-Base Processes: These processes unavoidably yield salt and water as by-products. The products of redox reactions are oxidized and reduced copies of the reactants. Water is a substance that is utilized routinely.
- Precipitation Reactions: These reactions result in the production of solid precipitates or suspensions. Synthesis methods generate molecules that are bigger than the reactants. These chemical reactions create products with less potential energy than the reactants.
- Endothermic Reactions: These generate products with greater potential energy than the reactants. Reactants are the first elements of a chemical reaction. They are distinct from the reaction's results. In every chemical reaction, there are reactants and products. The reactants of a combustion process of the candle are candle wax and the oxygen in air. Water vapor and carbon dioxide are the by-products. The reactants of methane gas combustion are methane (CH_4) and oxygen from the air (O_2). Carbon dioxide (CO_2) and water(H_2O).

are generated as a result of this process Hydrogen (H_2) and oxygen (O_2) are the reactants involved in the creation of water (Ebbesen, 2016). Water is the ultimate consequence (H_2O). The photosynthetic reactants are carbon dioxide (CO_2) and water (H_2O). Glucose is the substance of concern (C_6H2O6). It is vital to realize that sunlight is not a reactant.

Reactants are material as opposed to energy (atoms, molecules, and ions). In chemical equations, it is vital to separate reactants and products. Examine the reaction arrow in a chemical equation to discover the reactants and products. In a reaction that can only move forward, the arrow points from left to right. On the left side of the arrow are the reactants, while the products are on the right. A chemical species that appear on both sides of a reaction (such as solvent or spectator ions) is neither a reactant nor a product. A and B are reactants in the following reaction, and C is the product: C Equals A + B Nonetheless, there need not be more than one reactant. The reactant in this chemical reaction is A, and the products are B and C: A B + C In a balanced chemical equation, the products and reactants have the same number and sort of atoms. As an instance, the amount of hydrogen and oxygen atoms in the reactants (H_2 and O_2) and the product (H_2 and O_2) are similar (H_2O)

Figure 4.3: Laboratory testing of chemical reactants.

Source: By Elrond - Own work, CC BY-SA 4.0, https://commons.wikimedia.org/w/index.php?curid=81536340

What is the distinction between a reactant and a product, and how do they differ? The Equations, Symbols and equations in a chemical equation represent the numerous elements and molecules in a chemical process. It is a graphical depiction of a chemical process based on a chemical formula. In other words, reactants become products (or reactants become products). This can be expressed in two ways: "———->" and "=". Reactants and products are the chemicals that cause a chemical reaction to occur (Ebbesen, 2016). The acronyms below reflect the various levels of responder engagement. G denotes gas, L denotes liquid, S denotes solid, and Aq denotes an aqueous solution. 2Na(s) + $2H_2O(l)$ equals $2NaOH_{aq)}$ + $H_{2(g)}$ Stoichiometric coefficients (coefficients) are numbers that are added in front of formulations to reflect the reactant and product mixing ratios to balance a chemical equation.

The law of mass conservation applies to balanced equations. For example, the total mass of a reaction before and after the reaction is equal. This affects the number of atoms involved. It makes no difference whether the reaction is complete or not if there are still atoms involved. The following describes how sodium metal and chlorine gas interact: 2 NaCl(s) = 2 NaCl(s) + 2 NaCl(s) + 2 NaCl(s) + 2 NaCl(s) + 2 NaCl(s) + 2 NaCl(s) + 2 NaCl(s) + 2 NaCl(s) + 2 NaCl(s) + 2 NaCl(s) + 2 NaCl(s) + 2 NaCl(s) + 2 (s) For each size, the relative amounts are produced by multiplying the equation by a factor. There can be up to two of each of the following elements: Na, Cl_2, and NaCl (Na). This formula equals 4 Na(s) + 2 Cl_2(g) (s) This reaction may be seen on at least three different scales: When a chlorine molecule comes into touch with two sodium atoms, the sodium chloride formula unit is formed (nanoscale or atomic level). On a molar basis, two sodium atoms and one chlorine molecule are joined to produce two moles of sodium chloride. 45.98 grams of sodium and 70.90 grams of chlorine are mixed in a big reactor to produce 116.88 grams of sodium chloride (practical scale).

Stoichiometry is a mathematical formula that relates the reactant and product masses. The formula's equivalent quantities are stoichiometricconfigurations of Chemical reactions A third material is generated when two chemicals combine. A and B ----> AB The inclusion of two NaCl (s) and two CaO_3 (s) results in the formation of two NaCl (s) and two CaO_3(s) (s) A chemical decomposes into two or more new compounds throughout the breakdown process. When two things are combined, they produce two results. (s) 2Hg(l) + O_2(g) Compounds can be replaced (displaced) when one element interacts with another to generate a new compound. ——> 3 O2(g) + 2 kclo3(s) + 2 kcl(s) (g) (also known as a single replacement reaction). C + AB ——> AB When Fecl2 (aq) and H2 (g) (g) react, two agno3(aq) and one Cu(s) are produced (s) Cu(NO3) is a compound as an element (NO3) Each of the following elements has two atoms: 2(aq) and 2(s) Ag (s) A metathesis reaction is a chemical process that seems to include the exchange of elements between two reactants (also known as a double-replacement reaction) (Vasapollo et al., 2011). These interactions between ions in solution can result in the formation of insoluble compounds, weak electrolytes, and nonelectrolytes. To put it another way, Bacl2(aq) + Zncl2(aq) = Baso4(s) + Zncl2(aq) (aq) NaCl(aq) = NaCl(aq) + ZnS in this equation (s) The combustion reaction is another common process. When a chemical combine with oxygen, it produces a flame and a quick release of heat, which is known as combustion. When 2 C4H10 (g) and 13 O2 (g) are combined, they provide 8 CO2 (g) and 10 H2O (g)

Figure 4.4: Geneticist Riin Tamm wearing protective lab coat.

Source: By Lauri Kulpsoo - Own work, CC BY-SA 3.0, https://commons.wikimedia.org/w/index.php?curid=22487838

Balancing is required in all chemical equations. When attempting to balance an equation, trajectory and error is critical. The principles and ideas presented here may assist you in avoiding blunders. Reading the equation from left to right is required. The reactants are on the left side of the arrow, while the products are on the right. Two H_2O reactants are created as a result of two H_2 + O_2 reaction products. Rather than altering subscripts inside formulae, equilibrium is attained by tweaking the coefficients preceding formulations. If there is no coefficient, the formula assumes the value is 1. The best approach is, to begin with, an element that appears just once in a compound. Now, to the right or left, shift any constituent that appears in several compounds. Finally, balance the free elements. Any element that exists in the elemental form to the right or left of the centre must be balanced. Compounds or ions, for example, which comprise polyatomic species (such as H^+ and SO_4^{-2} on opposite sides of the arrow) should be balanced rather than as individual elements. If possible, utilize coefficients with the least whole-number ratio. As an example, $N_2 + H_2O + CO_2 = C_2H_8N_2$ + N2O4 to begin, balance either side of the arrow with any component contained in just one compound (Turney et al., 2012). The carbon and hydrogen balance will be re-established throughout this procedure. Because there are two C atoms on the left, CO_2 is preceded by a 2.

Each of the eight hydrogen atoms requires four molecules of H_2O: $C_2H_8N_2 + N_2O_4 = N_2 + 4\ H_2O + 2\ CO_2$ In the second stage, look at each element that appears in many compounds and decide whether it fits on the right or left side of the balancing equation. Because Step 3 dealt with nitrogen balance, this step just deals with oxygen balancing. Because there are four oxygen atoms on the left and right on the right, we add a 2 to N_2O_4. To make $C_2H_8N_2$ from N_2, combine 2 N_2O and 4 H_2O. 4 Then, move any newly appearing elements to the right or left of where they were previously. By putting a 3 in front of N_2 on the right, the nitrogen atoms can be balanced: In this example, 3 N2, $4H_2O$ and $2CO_2$ combine to form $C_1H_8N_2$, $C_2H_8N_2$, and $C_1H_8N_2$. Reactants (elements or compounds) are converted into products during a chemical reaction (compounds or elements). Chemical processes, on the other hand, are incapable of converting one element into another. Chemical interactions, on the other hand, result in the formation of a new substance.

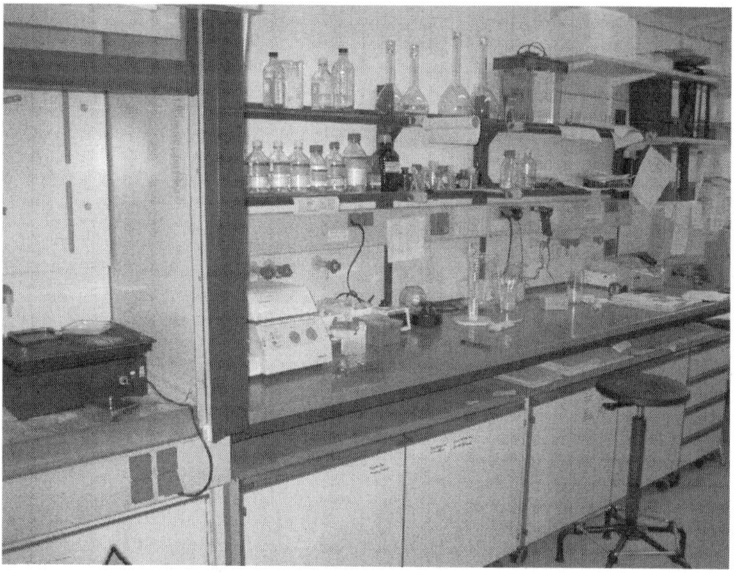

Figure 4.5: Laboratory bench.

Source: https://upload.wikimedia.org/wikipedia/commons/5/5f/Lab_bench.jpg

4.2 CHEMICAL REACTION SIGNS

There are various signs that a chemical reaction has occurred, including the following: Heat can be emitted or absorbed. Something fresh is seen with the naked sight. Chemical equations are used to represent reactions, products, and other factors such as energy changes and catalysts. Arrows in these equations represent chemical processes. A chemical reaction is frequently structured as follows: Reactants→Products

Consider what occurs when you switch on your natural gas cooktops. When methane (natural gas) combines with oxygen in the air, carbon dioxide and water vapor are formed (Antoni et al., 2015). The chemical equation for this reaction is shown in the graphic below: The equation looks like this: When one molecule of methane gas and two atoms of oxygen gas combine, one molecule of carbon dioxide gas and two atoms of water vapor are formed. The two numbers in front of the oxygen gas and the two numbers in front of the water vapor reflect the reaction coefficients. The number of chemical species provided indicates the number of chemical species that react or form.

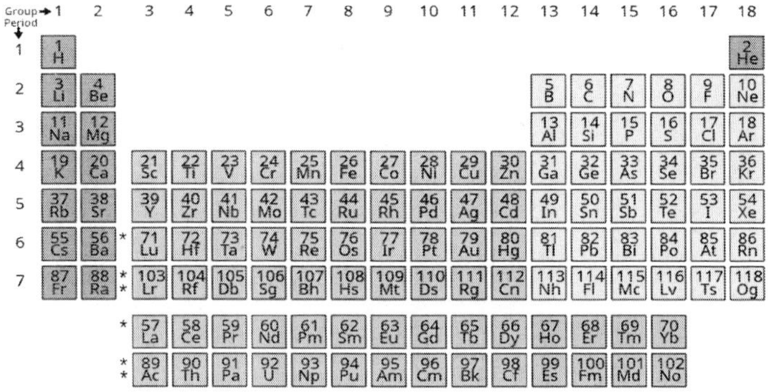

Figure 4.6: Chemical reactants on the elements table.

Source: By User:Double sharp, based on File:Simple Periodic Table Chart-en. svg by User:Offnfopt - Own work, CC BY-SA 4.0, https://commons.wikimedia. org/w/index.php?curid=98955505

The reactants are methane and diatomic oxygen, while the products are carbon dioxide and water. All reactants and products are gases, as shown by the g's in parenthesis. Throughout this method, all of the reactants and products are hidden from view. As evidenced by the heat released, a chemical reaction is taking place. This is a great example of an exothermic process, which produces heat. Exothermic reactions are widespread. Certain processes, however, require energy rather than releasing it. These reactions are described as endothermic. New compounds, known as products, are generated as the atoms in one or more reactants rearrange. A chemical reaction system is made up of two sorts of elements: reactants and products. We use the term "Reactants and Products" to define a chemical reaction system. To determine whether or not a chemical reaction represented by a balanced equation is conceivable, the first question must be answered. In theory, the answer is "yes" if all of the reactants and products are life-sustaining chemicals (Thomford et al., 2018). The following factors must be considered: Is it, as far as I can tell, comprehensive? In other words, how much of the reactants are converted into the end product? In certain chemical reactions, all reactants are transformed into products, but in others, just a few products are produced. The fact that so many are in the center indicates that considerable levels of each component remain.

As you advance through the course, you'll discover that you can use thermodynamic principles to forecast the possibility of a reaction occurring simply by looking at its parts. What is the reaction rate? One reaction may take milliseconds, while another may take years. The rate of a chemical reaction can vary substantially depending on temperature, matter state (gas, liquid, or solid), and the presence of a catalyst. Completeness and reaction time cannot be measured in the same way. What is the fundamental mechanism of the reaction? What occurs when reactants undergo atomic or molecular transformation into products? Is there a function for intermediate species (those that are produced but then consumed and hence do not show in the net equation)? To comprehend chemical change at the microscopic or kinetic level, tests must be carried out and the results are analysed. To be deemed kinetically restricted, the reaction must have a mechanism that permits it to continue swiftly. However, from a thermodynamic standpoint, a process that occurs swiftly but to a restricted degree is confined. As you'll see in the next section, industrial chemistry relies on a range of approaches to overcome these two sorts of bondts. Equilibrium. New compounds, known as products, are generated as the atoms in one or more reactants rearrange.

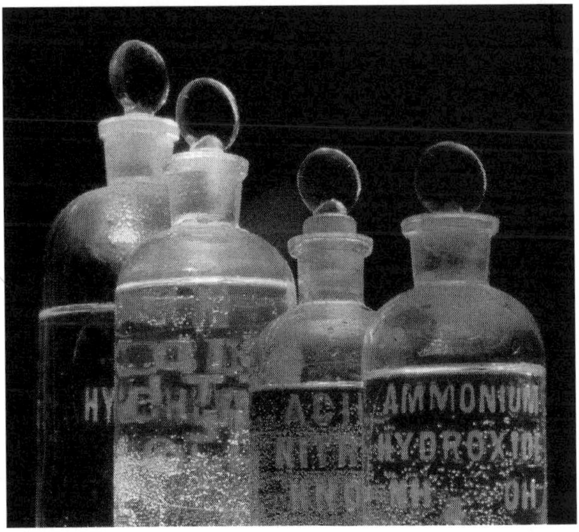

Figure 4.7: Chemical reactant-products.

Source: By Joe Sullivan - Flickr, CC BY 2.0, https://commons.wikimedia.org/w/index.php?curid=2167690

A chemical reaction system is made up of two sorts of elements: reactants and products. We use the term "Reactants and Products" to define a chemical reaction system. To determine whether or not a chemical reaction represented by a balanced equation is conceivable, the first question must be answered. In theory, the answer is "yes" if all of the reactants and products are life-sustaining chemicals. However, there are a couple of limitations to this reaction. What is the reaction's scope? In other words, how much of the reactants are converted into the end product? In certain chemical reactions, all reactants are transformed into products, but in others, just a few products are produced. The fact that so many are in the center indicates that considerable levels of each component remain (Aramwit et al., 2010).

As you advance through the course, you'll discover that you can use thermodynamic principles to forecast the possibility of a reaction occurring simply by looking at its parts. What is the reaction rate? One reaction may take milliseconds, while another may take years. The rate of a chemical reaction can vary substantially depending on temperature, matter state (gas, liquid, or solid), and the presence of a catalyst. Completeness and reaction time cannot be measured in the same way. What is the fundamental mechanism of the reaction? What occurs when reactants undergo atomic or molecular transformation into products? Is there a function for intermediate species (those that are produced but then consumed and hence do not show in the net equation)? To comprehend chemical change at the microscopic or kinetic level, tests must be carried out and the results should be analysed. To be deemed kinetically restricted, the reaction must have a mechanism that permits it to continue swiftly. However, from a thermodynamic standpoint, a process that occurs swiftly but to a restricted degree is confined. As you'll see in the next section, industrial chemistry relies on a range of approaches to overcome these two sorts of bonds. Equilibrium. The word "force balance" really refers to a "force balance." Mechanical equilibrium is defined in this context as a situation in which the forces acting on each other are balanced (Shin et al., 2019). The attraction between atoms that prevents two items from occupying the same space at the same time, in this case, the book's surface and the table's surface, entirely counteracts Earth's gravitational pull on the book's mass (its "weight").

Figure 4.8: Rust on a car due to chemical reaction.

Source: By This file is published under the following Creative Commons license:Attribution NonCommercial Unported 3.0 - Own work, GFDL 1.2, https://commons.wikimedia.org/w/index.php?curid=136219

When you pick up a book and raise it over the surface of the table, your arm generates an extra upward force that throws the balance off. If you want to maintain the book balanced above the table, adjust the upward force until the weight is properly balanced, restoring equilibrium.

4.3 MECHANICAL EQUILIBRIUM

Mechanical equilibrium exists when an object is either stationary or continually moving. Because the formula states that if an object's net force equals zero, its acceleration equals zero, it is conceivable to argue that it is in mechanical equilibrium if its velocity does not vary. A water molecule is made up of two hydrogen atoms and one oxygen atom.." To form two water molecules, we'll need four hydrogen atoms and two oxygen atoms. Ten hydrogen atoms and five oxygen atoms are required to make five water molecules. Any number of water molecules requires the same number of atoms: There are two hydrogen atoms for every oxygen atom. No matter how many water molecules are generated, hydrogen to oxygen atoms is consumed in a 2:1 ratio. Arranging atoms one by one, on the other hand, is extremely difficult, if not impossible. It was once remarked that we deal with billions of atoms at the same time (Shi et al., 2010). How will we handle so many atoms (and molecules) at once? Instead of counting atoms one by one, we use mass to do this operation. 1 u is the approximate mass of hydrogen atoms. The mass of an oxygen atom is about 16 u. The oxygen-to-hydrogen mass ratio is around 16:1. In other words, the mass ratio of an element with two atoms is 32:2, which is equivalent to 16:1. With each element having 12 atoms, the mass ratio is about (12 16)/12 (11):(12 1), or 192:12. The mass ratio of 100 atoms is 1,600:100, or 16:1.

Figure 4.9: Chemical reactant in a reaction bottle.

Source:https://media.istockphoto.com/photos/automated-chemical-synthesis-picture-id1145857171?k=20&m=1145857171&s=612x612&w=0&h=cajzR22hW4LxpEiOUKjCaEZ3ljQq6QSVh-eK9pE3rOg=

If there are equal amounts of hydrogen and oxygen atoms in a solution, the ratio is always 16:1. When the mass ratios of other elements are compared, they show the same regularity. Although silicon atoms are 28:1 heavier than hydrogen atoms, calcium atoms are 40:7 heavier than lithium atoms. As a result, we've shown that atom masses are constant concerning one another as long as the amount of each sort of atom remains constant. Consider a larger-scale scenario. You will always have the same number of Ca or Li no matter how much you have. To connect macroscopic amounts of objects, we need a number that represents an adequate number of atoms. Because of their tiny size, even 12 atoms cannot be used (Aidas et al., 2014).

There must be a way to represent billions upon billions of atoms. The term "mole" is used in chemistry because it is defined as 6,02 1023 pieces, which is a huge number of atoms or molecules. In terms of the number of items it can carry, a mole (abbreviated mol) is similar to a dozen. The number 6.02×10^{23} is used to represent macroscopic atom and molecule concentrations. It is named after the 19_{th}-century physicist Avogadro.

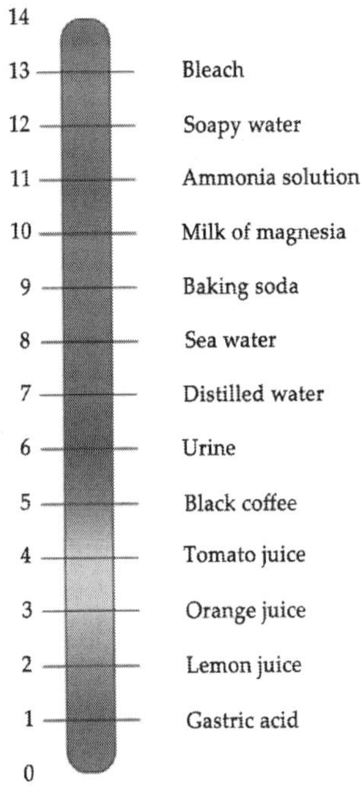

Figure 4.10: Image of the pH scale.

Source: https://en.wikipedia.org/wiki/File:PH_Scale.svg

One mole of oxygen has 6,02 1023 6.02×10^{23} oxygen atoms. There are 1.2044 10^{24} Na atoms in 2 mol of Na atoms, or 6.02×10^{23}Na atoms. A mole of benzene (C_6H_6) equals 0.5 (6.02×10^{23}) or 3.011 1023 molecules of C_6H_6. A wide range of chemical entities may be described using the molecular unit. In these circumstances, we employed moles of atoms and moles of molecules. The name "mole" can refer to any of the 6.02×10^{23}different species. It does not, however, indicate which species it refers to. Examples include atoms, formula units (such as ionic compounds), and molecules. These details must be kept up to date regularly. Because each H_2 molecule contains two H atoms, one mol of H_2 molecules (6.02×10^{23}molecules) contains two mol of H atoms. Calculations based on the number of atoms in each chemical can assist us in connecting molecular numbers to molecular atom numbers. From a single molecule of ethanol, the following relationships may be deduced: (C_2H_6O). For instance, a molecular reaction or a chemical reaction (Sneskov & Christiansen, 2012). A balanced equation can be used to compute the molar ratios of the reaction participants. Butane, C4H10, has the following balanced combustion equation: 8 CO_2(g) + 10 H_2O = C_4H_{10}(g) (g) This equation may be used to infer the following relationships: Several formulae might be used: 13 moles O_2/2- mol C_4H_{10}, 8 mol CO_2/2mol C4H10C_4H_{10}, 13 moles O_2/2- mol C_4H_{10}, or 13 moles O_2/2- mol C_4H_{10}.

These ratios allow you to calculate the number of one participant depending on the number of another. Example How much oxygen is produced when 0.50 mol of butane combines with it? 2 C_4H_{10}(g) + 13 O_2(g) = 3.25 mol O_2 = 8 CO_2(g) + 10 H_2O(g) (0.50 mol C_4H_{10}) Example The interaction between alkali metals and water produces hydrogen gas as well as alkali metal hydroxide. This is an illustration of a lithium-water reaction: Each litre of a solution of 2Li(s) and 2H_2O(l) contains two litres of water (g) How many moles of H_2 are produced when Li fully interacts with water? In 3.12 mol, there are comparable quantities of Li and H2H_2 for every mole of H_2 in the system. Molar masses and molar ratios can be used to calculate the mass of any other participant in a chemical equation (Stuppia et al., 2012). To begin, calculate the chemical's molecular pressure in grams. Using a suitable molar ratio, convert the moles of a reactant or product to the moles of another participant. The final step is to convert the second participant's molecular pressure to grams. Example: How many grams of H_2 are created when 80.57 grams of Li completely interact with water? Each liter of a solution of 2Li(s) and 2H_2O(l) contains two litres of water (g) 1 mol Li/6.941 g Li = 1 mol Li (80.57 g Li) The following is the second step: (11.6078 moles of lithium) = 5.8039 moles of hydrogen the third and last phase is as follows: A mole of hydrogen contains roughly 11.72 grams (1.72 moles) of H2 (5.8039 mol H_2) Example: Our bodies break down or degrade the food we ingest to produce energy for growth and development.

Figure 4.11: Chemical equipment in chemistry analysis.

Source: By Tweenk - Own work, CC BY 3.0, https://commons.wikimedia.org/w/index.php?curid=10587531

The following equation is used to convert glycogen ($C_6H_{12}O_6$) into carbon dioxide (CO_2) and water: $C_6H_{12}O_6$ +$C_6H_{12}O_6$6 + 6O_2 = $C_6H_{12}O_6$ + $C_6H_{12}O_6$ + 6O_2. When human drinks 856 grams of $C_6H_{12}O$over time, how much CO_2 is produced? The formula is as follows: $C_6H_{12}O$, with a mass of 856 grams, divided by the mass of $C_6H_{12}O$, with a mass of 180.2 grams (4.7503 mol $C_6H_{12}O$) Next: (6.75 moles CO_2/1 mole $C_6H_{12}O$) (4.7503 moles $C_6H_{12}O$) (28.502 mol CO_2)(44.01 g CO_2/1 mole Co_2)= (1.25 x 103 g CO_2) Example Pt(NH_3)2C_{12} (Pt(NH_3)2C_{12}) has long been used as a cancer therapy. This chemical is made using K_2PtptCl_{14} and ammonia in the production process (NH_3). Pt is obtained from aqueous solutions of 2NH_3 and K_2PtCl_4(NH_3) To arrive at Pt, we must apply the following formula: 2Cl_2(s) + 2kCl (NH_3) = Pt (aq) How much ammonia is required to react 100 grams of k2ptcl4K_2PtCl_4 with ammonia? 0.4818 moles of ammonia = 0.2409 moles of K_2PtCl_4 per kilogram of K_2PtCl_4(0.2409 mol K_2PtCl_4) The last step is to: 0.4818 mol N_3 = 8210 g N_3/mol NH_3 Increasing the Controllability of Reagents: A balanced equation depicts reactions and products in equal proportions. Even though this connection is constant, the number of reactants may vary. At any time, one or more of the reactants may be in excess. In other words, the number of products produced is proportional to the amount of a single reactant present.

Figure 4.12: The chemical laboratory testing.

Source: https://commons.wikimedia.org/wiki/File:Chemistry_Lab.jpg

The following are methods for determining product mass when one or more reactants are present in a limiting mass: Determine the maximum yield of a single product for each of the reactants (in moles or grams, as applicable). The limiting reactant in the second stage is the one that produces the least amount of the final substance. The amount of limiting reactant present determines the reaction products. Example When 4.0 moles of H_2 and 1 mole of O_2 react as described in the reaction, H_2O is formed. When two molecules of hydrogen are combined with one molecule of oxygen, two molecules of water are formed (Rhoads et al., 2012). Because only 2.0 mol of water is produced in this reaction, adding two molecules of hydrogen to one molecule of oxygen yields two molecules of water (1.0 mol O2).

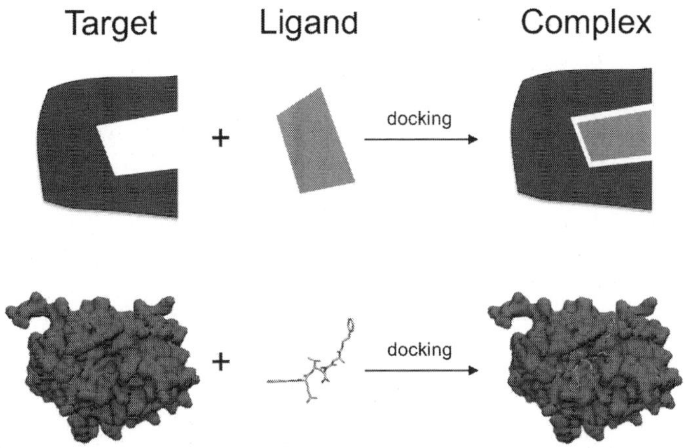

Figure 4.13: Chemical calculations on a board.

Source: By Scigenis - Own work, CC BY-SA 4.0, https://commons.wikimedia.org/w/index.php?curid=45515965

CHAPTER 5

CHEMICAL REACTIONS

CONTENTS

A chemical reaction occurs when one or more reactants are converted into one or more distinct products. Compounds and chemical elements are examples of substances. By rearranging the atoms of the reactants, chemical reactions produce a wide range of products. Chemical reactions are crucial in contemporary culture, technology, and even life itself. Fuel combustion, iron smelting, glass, and pottery manufacture, beer production, and the creation of wine and cheese are some examples of chemical processes that have been employed for millennia. In addition to the chemical reactions that occur in the earth's crust, atmosphere, and seas, several complex processes occur in all living systems. It is necessary to distinguish between chemical reactions and physical changes. Physical transitions include ice melting into water and water evaporating to create vapor (Rastogi & Sani, 2011). A substance's physical properties may change throughout metamorphosis, but its chemical identity does not. Water (H_2O) is the same substance regardless of its physical state since each molecule includes two hydrogen atoms and one oxygen atom. When water reacts with sodium metal (Na), it produces molecular hydrogen (H_2) and sodium hydroxide (NaOH). This denotes the occurrence of a chemical reaction or change. Understanding the fundamentals of chemical interactions.

Figure 5.1: Chemical analysis of chemical reactants.

Source: By No machine-readable author provided. Burn~commonswiki assumed (based on copyright claims). - No machine-readable source provided. Own work assumed (based on copyright claims)., CC BY-SA 3.0, https://commons.wikimedia.org/w/index.php?curid=298098

A chemical reaction occurs when two or more molecules combine to form a new substance (s). When reactants and products come into touch with one another, compounds are formed. Chemical interactions are important in many industries, communities, and even our daily lives. They are present in all of our activities, from iron rusting to pottery making to wine fermentation. Chemical reactions include physical changes such as precipitation, heat generation, color shifts, and so on. When two atoms, ions, or molecules combine to form a new link without destroying or producing an atom in the process of creating a new product out of the reactants, a chemical reaction occurs. Calculations in the scientific field: With so many chemical processes all around us, a nomenclature was developed to help understand chemical equation representations of reactions. A chemical equation in mathematics expresses how reactants are converted into products as well as the conditions under which that conversion occurs.

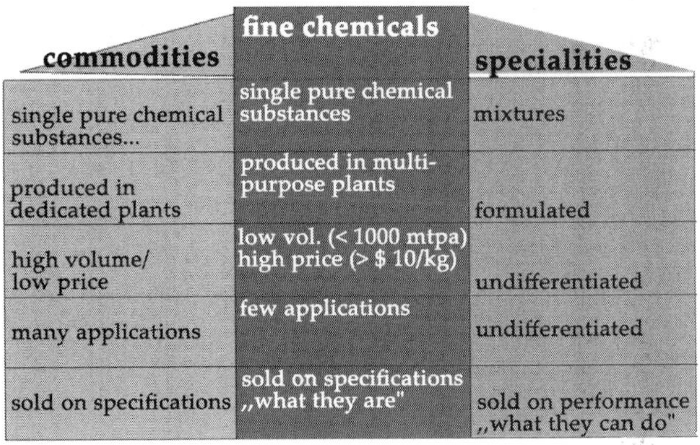

Figure 5.2: Chemical calculations on a board.

Source: By PeterRPollak - Own work, CC BY-SA 3.0, https://commons.wikimedia.org/w/index.php?curid=19652814

5.1 HISTORY OF CHEMICAL REACTIONS

The development of this theory may have been influenced by studies that classified chemicals as elements and compounds, as well as hypotheses describing how these processes worked. Without the concept of a chemical reaction, modern chemistry would not be possible. The first significant study in this discipline focused on gases. The discovery of oxygen by Carl Wilhelm Scheele and Joseph Priestley in the 18th century was significant.

Antoine-Laurent Lavoisier, a French chemist, made important contributions to the quantification of chemical processes with his theories. According to Lavoisier, elements gained weight as they oxidized, and he also discovered the link between oxygen interaction and element weight gain. He demonstrated the concept of chemical reactions using elemental combinations, inspiring others to study experimental chemistry quantitatively (Raies &Bajic, 2016). The development of the atomic theory was a significant milestone in the development of chemical processes. Dalton, an early-nineteenth-century English scientist, is largely responsible for proposing the atomic idea. Everything, according to Dalton, is made up of tiny, indivisible pieces known as atoms.

These atoms are specific to each element and rearranging them to form new compounds necessitates the use of chemical processes. In light of the present issue, this viewpoint on chemical processes is relevant. Understanding Dalton's theory established the framework for understanding earlier findings, such as the law of constant composition (elemental compositions are the same in all samples of material) and the rule of conservation, which states that matter cannot be generated or destroyed (all samples of a substance have identical elemental compositions). This is how we came to understand chemical processes, which are at the core of modern chemistry, by fusing experimental data with theoretical predictions. Theoretical chemistry helps illuminate the importance of today's experimental chemistry's many instances.

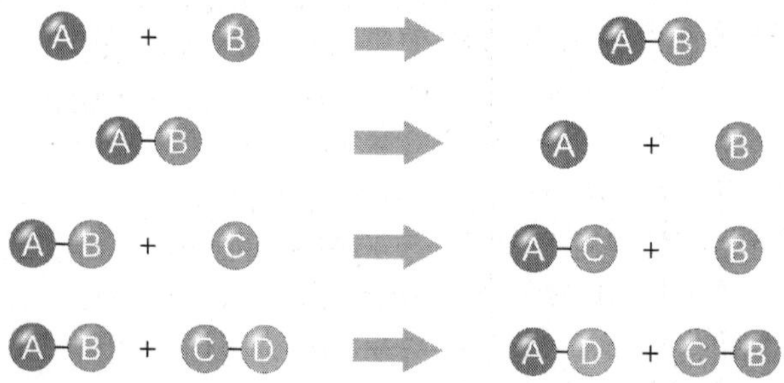

Figure 5.3: Chemical reactants.

Source: By Daniele Pugliesi - Own work, CC BY-SA 3.0, https://commons.wikimedia.org/w/index.php?curid=20643389

5.2 CHEMICAL REACTIONS OF VARIOUS TYPES

In a wide spectrum of chemical reactions, reactants, products, modifications, and other variables all play a part.

The following are some of the reactions:

- Reaction to combustion
- The decomposition procedure This is an ineffective answer. In reverse order, carbon dioxide and oxygen are oxidized. A synthesis process
- The reaction of combustion

When a flammable substance is combined with an oxidant, the result is an oxidized product. An oxidizer is a chemical that assists in the combustion process when a fuel is burned. One example is the burning of magnesium. When two magnesium atoms react with a single oxygen molecule, two molecules of magnesium oxide and heat are formed. A method of decomposing organic materials A single substance gets broken down into a group of smaller ones during decomposition. Breaking the molecule's bonds requires certain changes in the energy of the environment, such as heat, light, or electricity. Consider the transformation of calcium carbonate into CaO (quick lime), a crucial component of cement (Arif et al., 2010). The attack's antidote is a neutralization reaction, an acid, and a base mix to generate salt and water. The water molecule is made up of OH– ions and H+ ions.

When a strong acid and a strong base are neutralized, the resulting pH is 7. The neutralization of sodium chloride (common salt), water, and hydrochloric acid/sodium hydroxide yields sodium chloride and water. In the fourth oxidation-reduction reaction, electrons are moved from one chemical species to another in a reduction-oxidation reaction. Consider a zinc-hydrogen redox reaction in an electrochemical cell. Electrons are transferred from the zinc atom to the positively charged hydrogen ions when it comes into contact with them, resulting in the creation of a zinc ion-containing, long-lasting molecule. Reactions such as precipitation and double-displacement. Displacement reactions occur when anions and cations of two distinct compounds shift positions, resulting in the production of two new compounds. As an illustration, consider the interaction between silver nitrate and salt chloride. Silver chloride and sodium nitrate are the end products of the two-fold displacement procedure. It is one of the most fundamental types of chemical reactions, in which multiple simple molecules mix under certain physical circumstances to form a complex outcome (Pritchard et al., 2019). The result is always a combination. As an example, have a look at the interaction between sodium and chlorine gas. Sodium chloride, or table salt, is formed when two sodium atoms combine with chlorine gas. Important information to keep in mind. While the former creates a new compound, the latter affects the state of the original substance.

Figure 5.4: A chemical reaction of chemical reactants.

Source: https://media.istockphoto.com/photos/close-up-of-chemical-reactions-in-beakers-at-laboratory-picture-id1217973957?k=20&m=1217973957&s=612x612&w=0&h=h5RYCDNWBzAayIDGnm44dducinrEI6Y9KCAfEeIYunE=

- The concept of mass conservation governs chemical processes. When a reaction occurs, new atoms or molecules are produced as a result of the reaction between the reactants and products. As a result of this process, neither new nor old atoms are created or destroyed. The fundamentals of chemistry lie at the heart of chemical processes. When creating a new chemical, chemists use the term "synthesis" or "synthesize" to describe it. A chemical equation depicts the process of converting raw materials into products. For example, iron sulfide (FeS) is formed when sulfur (S) and iron (Fe) mix Using the + sign, we can see that iron and sulfur interact with each other. FeS = Fe() + S. The arrow points to the formation of iron sulfide as a result of the reaction. In chemical reactions, the states of matter of the reactants and products are indicated by the symbols (s), (l), and (g) No matter how many times you do a routine scientific experiment, nothing changes.

Reaction equations must, therefore, have the same number of atoms from each class present on the two opposing sides. According to the iron-sulfur balancing equation, a single iron atom may combine with a single sulfur atom to generate a single formula unit of iron sulfide (FeS_2). In most cases, chemists work with quantifiable quantities of elements and compounds. A formula for iron-sulfur equations includes FeS, which represents 87.911 grams of iron sulfide, 55.845 grams of iron, and 32.066 grams of sulfur (Pinzi& Rastelli, 2019). During a chemical reaction, no matter is created or destroyed, hence the total mass of the reactants and products is unaltered. This reaction produces eight times as much iron sulfide as when only one-tenth the iron (5.585 grams) and one-tenth the sulfur (3.207 grams) are employed (8.791 grams). Sulfur would be reduced to 28.859 grams if 32.066 grams of sulfur and 5.585 grams of iron were present at the start of the operation.

Methane (CH4, a significant component of natural gas) combines with molecule oxygen (O_2) to produce carbon dioxide (CO_2) and water (H_2O) using the chemical equation $CH_4(g) + 2O_2(g)\ CO_2(g) + 2H_2O(l)$. Before oxygen and water, a stoichiometric factor is 2. There is a 1 before CH_4 and CO_2 to give you a hint. This means that when two molecules of oxygen interact with one molecule of methane, onc molecule of carbon dioxide and two molecules of water are formed. One carbon, four hydrogen, and four oxygen atoms on either side of each element ensures that the equation is balanced (here one carbon, four hydrogen, and four oxygen atoms) (here one carbon, four hydrogen, and four oxygen atoms). There is no difference in the amount of carbon dioxide and water produced by 16 grams of methane vs. 64 grams of oxygen.

As a consequence, using 80 grams of reactants yields 80 grams of finished goods. The stoichiometry of a chemical reaction is the connection between the reactants and the products. Stoichiometry uses the mole concept to determine mass bonds in chemical processes based on the principle of matter conservation. The number of atoms or molecules in any element or compound is the same. Carbon-12 has a molar mass of exactly 12 grams and contains $6.022140857 \times 10^{23}$ atoms (Avogadro's number) per mole, according to universal agreement. A mole of iron weighs 55,847 grams, but a mole of methane, molecular oxygen, and water each weigh 31,999 grams and 18,015 grams, respectively, in pressure. $6.022140857 \times 10^{23}$ molecules make up each mass.

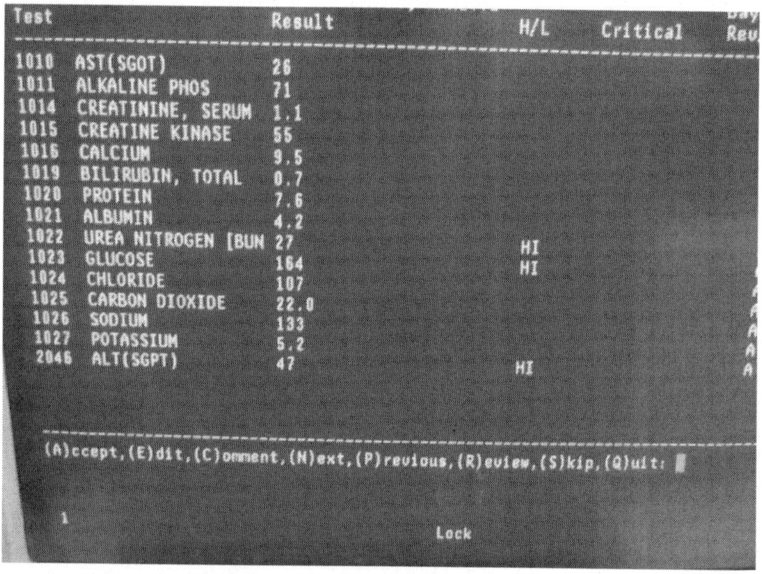

Figure 5.5: Chemical analysis of chemicals.

Source: By Bobjgalindo - Own work, CC BY-SA 4.0, https://commons.wikimedia.org/w/index.php?curid=5654948

Concerns concerning the application of power in chemical reactions, and energy are essential. As we presently understand them, chemical reactions need the breaking of bonds between atoms in reactants and the reassembling of atoms or sections of molecules into products by the formation of new ones. When bonds are dissolved, energy is absorbed; when bonds are created, energy is released. Breaking bonds requires more energy than making new ones, resulting in energy absorption. If the energy source is heated, the process is endothermic (Patil et al., 2008). Endothermic reactions are the polar opposite of exothermic processes. We use the words "exoergic" and "endoergic" to denote energy sources other than heat. Exothermic reactions occur in a variety of situations. Exothermic reactions are generally always used to generate compounds from their basic ingredients.

When calcium metal is mixed with oxygen gas, it produces a range of oxides that are unique from water, such as calcium oxide (CaO). The burning of fuels is one of the most well-known exothermic processes (such as the previously reported interaction of methane with oxygen). When water is added to lime, slaked lime (calcium hydroxide, $Ca(OH)_2$) is created (CaO). The reaction CaO(s) + H_20 (L) CaO_2 occurs in concrete, releasing energy as heat when the mixture warms. Exothermic (or exoergic) reactions do not always happen in the same way (or exoergic). Compounds that require energy input to form include NO and hydrazine. It is also an endothermic process that requires a high temperature to convert limestone ($CaCO_3$) to lime (CaO). When $CaCO_3$(s) and CO_2 are combined together, they equal $CaCO_3$(s) and CO_2 = $CaCO_3$(s) and CO_2 = $CaCO_3$(s) and $CaCO_3$ = $CaCO_3$(s) and CO_2 = $CaCO_3$(s) and CO_2 = $CaCO_3$(s) and CO_2 = (g).

Electrolysis, an endoergic mechanism, breaks down water into its atoms. Instead of using thermal energy, this method uses electrical energy. O_2 (g) andH_2O(g) are synonymous . When heat is created, reactions and products may more readily be transformed into one another (Barbatti et al., 2014). Only by considering the entropy can the favorable or unfavorable nature of a reaction be established. Entropy quantifies the number of different energy configurations that can exist in a system. Because of entropy, not all of the energy available in a process can be utilized to do work. If the sum of the entropy changes for the system and its surroundings is positive, a chemical process will yield more products.

The combustion of wood is a great example of this. Wood has a low value in terms of entropy. The byproducts of burning wood include ash, carbon dioxide gas, and water vapor, all of which have a high entropy value. During combustion, entropy increases. Furthermore, burning heat increases environmental entropy. Total entropy changes that benefit the product for the reactants and their surroundings suggest a positive reaction. Entropy reduces as hydrogen and oxygen combine to make water. In contrast, exothermic heat transfer increases the entropy of the surrounding environment, compensating for the loss. Because of the overall rise in entropy, fuel combustion with hydrogen is once again favored. Some Considerations in Kinematics The majority of chemical reactions begin with a burst of energy. Starting a fire with a burning match or spark requires the exothermic combustion of wood, paper, or methane. When a match is rubbed against a suitable surface, frictional heat causes an exothermic chemical reaction, releasing the stored energy in the match. Light can give the energy needed to launch various processes.

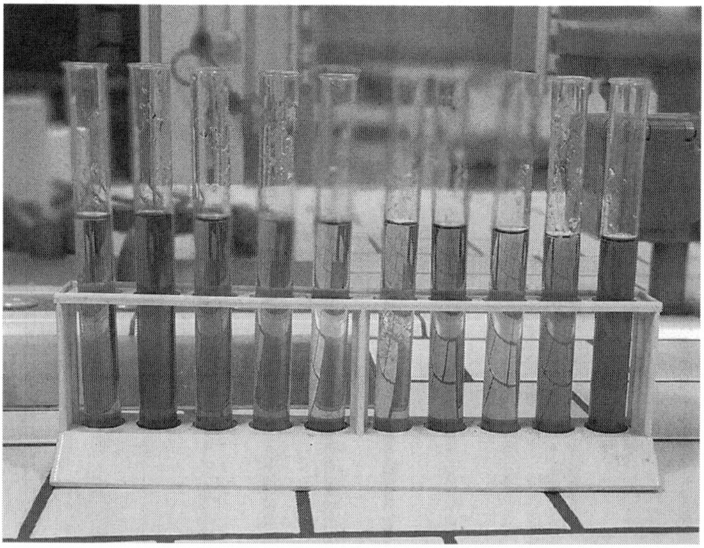

Figure 5.6: Chemical equipment in chemical analysis.

Source: By Alvy16 - Own work, CC BY 4.0, https://commons.wikimedia.org/w/index.php?curid=84933482

As a result of solar radiation, several photochemical, or light-driven, processes occur in the Earth's atmosphere. Photosynthesis in the troposphere converts ozone (O_3) to oxygen (O_2). The sun's ultraviolet (UV) light is absorbed by the Earth's atmosphere, preventing hazardous high-energy radiation from reaching the planet's surface. A favorable reaction doesn't need to occur. Reaction time must be evident as well. Factors such as reactant concentration, temperature, and the presence of catalysts all influence reaction rates. Concentration, which is necessary for all reactions, influences the collision rate of reacting molecules. Because reactions can only occur when two molecules with sufficient energy collide (Ogino et al., 2013). The number of molecules with enough energy to react is affected by temperature. Catalysts influence the rate of a process by providing a lower-energy pathway for it to take place.

Precious metal compounds are often employed as catalysts in car exhaust systems to convert hazardous pollutants such as nitrogen dioxide into innocuous nitrogen and oxygen. Two well-known examples of biological catalysts are enzymes and chlorophyll in plants (which allows atmospheric carbon dioxide to be transformed into complex organic compounds such as glucose). Pepsin, for example, assists in the digestion of big protein molecules.

Chemical reactions that are organized in a precise manner Chemists classify reactions in four ways: by product type, reactant type, reaction outcome, and reaction technique. Reactions are frequently classified into two or three kinds. The classification of items is determined by the type of product.

Gas production processes: A variety of processes produce carbon dioxide, hydrogen sulfide, ammonia, and sulfur dioxide (SO_2). Carbon dioxide is produced when hydrochloric acid (HCl) reacts with a metal carbonate, such as calcium carbonate ($CaCO_3$, a key component of limestone, seashells, and marble). 2 HCl(aq), $CaCl_2$, and CO_2 aqueous solution, H_20 + $CaCO_3$(s), and 2 HCl(aq), $CaCl_2$, and CO_2aqueous solution (l) The chemical present in an aqueous solution is denoted by the symbol (aq). The rise of cake batter is caused by an acid-baking soda reaction involving sodium hydrogen carbonate ($NaHCO_3$) (sodium bicarbonate, $NaHCO_3$). Tartaric acid ($C_4H_6O_6$) is an acidic reactant found in a variety of foods (O'Gorman et al., 2012). The aqueous solutions of $C_4H_6O_6$are CO_2 (g) and H2O (l), and the gaseous solutions are $C_4H_6O_3$ and $NaHCO_3$ (aq) (g) In this equation, it is written as $NaC_4H_50_6$. Most baking powders contain tartaric acid and sodium hydrogen carbonate, but they are separated by starch.

When baking powder is added to a wet dough, sodium hydrogen carbonate and citric acid partially dissolve, allowing them to react. The batter rises due to the presence of carbon dioxide. Precipitation processes: When two solutions with different types of anion and cation are mixed, an insoluble compound can form, as shown in the figure The solid that has been separated from the liquid is referred to as precipitate. Sulfur dioxide (SO_2), hydroxide (OH), carbonate [CO_3^{2-}], and phosphate (P) are some of the most common insoluble substances in water (PO_4^{3-}po43). If an anion such as Fe^{2+}, Cu^{2+}, or Al^{3+} is introduced into a solution that already contains one of these metal cations, precipitation will occur. Chemicals in water do not dissolve minerals. Several natural precipitation mechanisms, such as the formation of metal sulfides in so-called "black smokers," may explain mineral production.

The types of reactants used in a reaction are classified. There are two types of reactions that transfer charged species. Electrons are transferred from one reagent to another during oxidation-reduction processes. During acid-base reactions, protons (H^+) are exchanged between acids and bases in water. In reactions of oxidation and reduction, an electron is transferred from a reducing agent to an oxidizing agent in a redox reaction, usually one or more at a time. For example, in the reduced material, this decreases an atom's real or apparent charge while increasing it in the oxidizing substance. Simple redox reactions require an element's interaction with oxygen. Magnesium oxide (MgO), for example, is formed when magnesium is burned in the presence of oxygen (MgO).

As a result, we end up with an ionic molecule composed of Mg^2 and O^2 ions. In this process, two electrons are lost from each magnesium atom and two electrons are gained from each oxygen atom. Because of the redox reaction, iron rusts when exposed to moisture. 2Fe(OH) is equal to 2Fe(s) + $2H_20(l)$ + $O_2(g)$ (oh) To produce iron dihydroxide (Fe(OH), ferrous metal is oxidized with elemental oxygen (o2). Redox processes are how batteries generate energy. A battery's electric current is formed when electrons are moved from a reducing agent to an oxidizing agent via an external circuit (Neese, 2012). In a conventional dry cell and alkaline battery, two electrons are delivered to the oxidizing agent per zinc atom, resulting in the transformation of zinc metal into the Zn^{2+} ion. Dry-cell batteries (commonly used in flashlights) collect electrons generated by zinc by using ammonium ions (NH_4^+) in the form of ammonium chloride (NH_4Cl. In alkaline batteries used in calculators and observes, electrons are transported to a metal oxide, such as silver oxide (AgO), and then reduced to silver metal. Any description of chemical processes would be incomplete if it did not include the ubiquitous acids and bases.

Figure 5.7: Image of chemical products.

Sources: By Kuebi = Armin Kübelbeck - Own work, CC BY-SA 3.0, https://commons.wikimedia.org/w/index.php?curid=6000765

5.3 SEVERAL THEORIES EXIST TO EXPLAIN ACID-BASE REACTIONS

In the Arrhenius theory, to put it another way, an acid increases the concentration of the hydronium ion H_3O^+in solution, whereas a base increases the concentration of the hydroxide ion (H_2Oin solution (OH). Acids such as sulfuric acid (H_2SO_4), nitric acid (HNO_3), and acetic acid (CH_3COOH) are also well-known (CH_3COOH). Common bases include sodium hydroxide (caustic soda) and slaked lime (lime soda) (calcium hydroxide) When water and ammonia (NH3) react, the result is a basic solution, as shown in the balanced equation below. Water as well as NH_3(aq) (l) NH_4^{+n}h4+(aq) + OH (aq) NH_4^+aq) + OH (aq) In this case,NH_4^+(aq) and OH (aq) are used (aq) While there is little to no significant increase in hydroxide ions, the reaction occurs.) Despite the low concentration of hydroxide ions, this reaction occurs (Nwodo et al., 2012). Morphine, cocaine, nicotine, and caffeine are among the many natural and synthetic bases available. These compounds, like ammonia, react with water to produce a solution containing the hydroxide ion.

Amino acids can act as both acids and bases. Acidic and basic sites are found in amino acid molecules. Aqueous solutions contain the molecular and "zwitterionic" forms of amino acids, h3n + ch2co2. The oxygen atom in the acid group is negatively charged, whereas the nitrogen atom is positively charged. According to the Arrhenius theory of acid-base processes, hydrogen ions (H^+) and hydroxide ions (OH^-) interact to form water. The interaction of sodium hydroxide and hydrochloric acid in water is one example. water (l) and NaCl (aq) (l) In addition to the other ingredients, HCl is required to create NaOH (l). Bronsted and Lowry developed a theory in the Bronsted-Lowry hypothesis. The Bronsted -lowry theory was developed by physicist Johannes Nicolaus Bronsted and English chemist Thomas Martin Lowry, in which acid is a proton giver and the base is a proton acceptor.

In this paradigm, acid-base reactions are viewed as balanced. (1) Base + (1) acid (1) acid combined with an acid (2) Two acids and one acid (2) The products are capable of dynamically reforming their reactants. The acids and bases (1) combine to form a conjugate acid-base pair. This theory has the advantage of being able to predict the future. The relative strengths of acid and base determine whether the equilibrium is in favor of the reactants or the products (product-favored). Water and the Bronsted-Lowry theory are frequently associated. When an acid is dissolved in water, it produces the hydronium ion and the acid anion (Moseti et al., 2016). The acid's strength is affected by whether the equilibrium favors reactants or products. Hydrochloric acid is one of the most powerful acids known to man, completely ionizing in water to produce hydronium and chlorine (Cl). We can assume that (aq) Hexane (aq) + water in the absence of any other information (aq) According to the Brnsted-Lowry hypothesis, ammonia, and hydrochloric acid react in water as follows: The following equations are simple to understand: NH4++ + Cl = NH_3(aq) + HCl(aq) Ammonium ion and ammonia make up this conjugate acid-base pair. This chemical reaction occurs when an acid (HCl) liberates a hydrogen ion from a hydrogen ion in a base (NH3). In this case, the equilibrium products benefit the weaker acids and bases. The Arrhenius and Bronsted-Lowry hypotheses differ in that a hydroxide ion is not included in this equation.

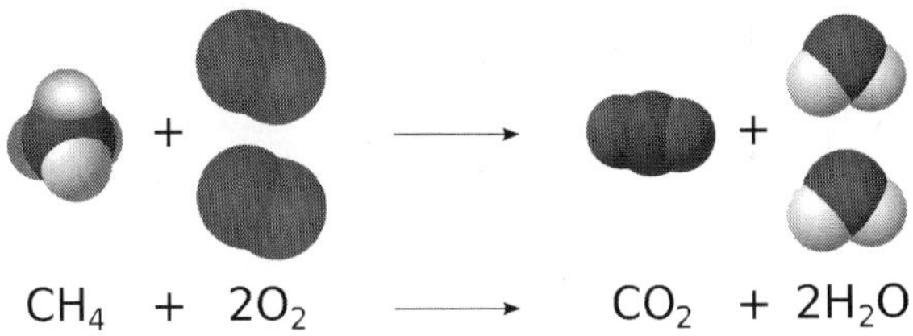

Figure 5.8: Combustion.

Source: By Yulo1985 - Methane-3D-space-filling.svgFile:Oxygen_molecule.svg Carbon-dioxide-3D-vdW.svgWater molecule 3D.svg, CC BY-SA 4.0, https://commons.wikimedia.org/w/index.php?curid=93335870

According to the Lewis Hypothesis, Gilbert Newton Lewis, an American physicist, proposed a more general acid-base theory. Lewis' theory employs the terms, electron-pair donors and acceptors. The electron pairs of the Lewis acid and base are shared in acid-base reactions. Ammonia is a good example for a Lewis base. A pair of nitrogen atom electrons can form a chemical bond with a Lewis acid such as boron trifluoride (BF_3). In this equation, each electron pair is represented by a colon. Metal ions react with ammonia, water, and other Lewis bases to form coordination compounds, also known as coordination compounds. These species require an additional Lewis acid-base reaction to being produced. Because of the presence of the $[Cu(H_2O)_6]_{2+}$ ion, when Cu^{2+} ions are dispersed in water, their solution takes on a light blue hue.

When the ammonia molecules in this solution replace the water, molecules attached to the copper, the $[Cu(NH_3)_4]^{2+}$cu(nh3)4]2+ ion is formed. Analyzing reaction outcomes to determine categorization Chemical reactions are frequently classified based on the outcomes they produce. This section categorizes numerous reactions. Many reactions, as previously stated, are difficult to categorize and may fall into multiple categories (Bacellar et al., 2015). Decomposition reactions reduce the complexity of chemical species. It is not uncommon to add energy to decomposition reactions. To put it another way, the sum of $2kclO_3$ plus 2kCl plus $3O_2$ equals $2kClO_3$ plus 2kCl (g) Another decomposition reaction (Cl_2) The sum of the parts is as follows: 2NaCl (l) + cl2 = 2NaCl, and a half gram of Cl2 (g) Mercury oxide (HgO) decomposition by heat and oxygen gas is a significant decomposition reaction in chemical history. Substitution, elimination, and addition are examples of chemical reactions. These terms can be used to explain organic reactions. A substitution reaction occurs when one atom or group of atoms in a molecule is replaced by another atom or group of atoms. Chloromethane (CH_3Cl) is produced as a topical anesthetic when methane (CH_4) reacts with chlorine (Cl_2). In this reaction, a chlorine atom replaces a hydrogen atom. Substitution reactions are commonly used in industrial chemistry.

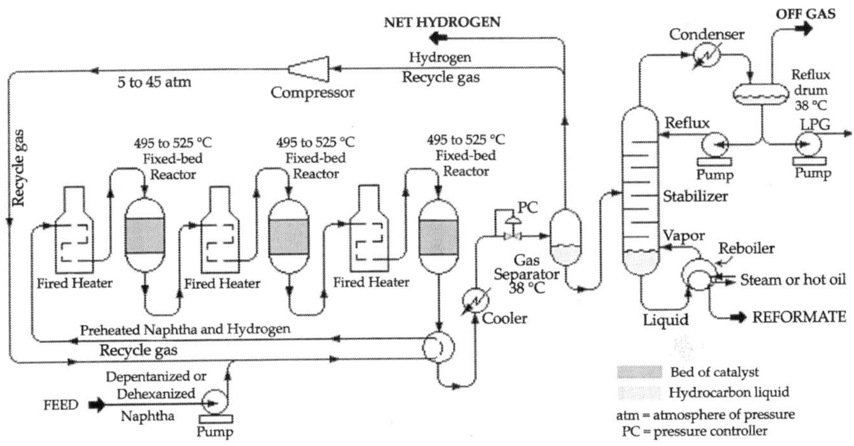

Figure 5.9: Chemical processes.

Source: By User:mbeychok - I drew this flow diagram myself and currently I own all rights to it. I used Microsoft's Paint program to draw it. I am User:mbeychok and the date is December 5, 2006., CC BY-SA 3.0, https://commons.wikimedia.org/w/index.php?curid=1435809

To form chlorodifluoromethane ($CHClF_2HCl$), two of the chlorine atoms in chloroform ($CHCl_3HCl$) are exchanged for fluorine atoms (CHClFHCl). This product undergoes a second reaction when heated to a high temperature. In this example of an elimination reaction, hydrogen and chlorine are removed from the starting material and replaced with hydrochloric acid (HCl) Tetrafluoroethylene, a precursor to the commercially available polymer Teflon is the other product. Addition reactions are the inverse of elimination reactions. As the name implies, adding one molecule to another. Example: a typical industrial ethanol preparation (CH_3CH_2OH). Historically, this compound was created through fermentation. It has been commercially produced since the early 1970s by combining water and ethylene. C_2H_4+ CH_3CH_2OHcand H_20h2o. P

olymerization Reactions: Many monomers combine to form polymers, which are large molecules (Milo et al., 2010). Polymers, in addition to natural and synthetic clothing fibers, include plastics that have changed society. Polymers can be created in two ways: (a) by combining small molecules in an addition reaction, or (b) by combining two molecules (of the same or different types) by removing a stable small molecule, such as water.

Condensation polymerization is a type of polymerization that combines addition and elimination reactions. Thousands of ethylene molecules are polymerized in the first type of chemical reaction. [CH_2CH_2 - CH_2CH_2] NH_2C =CH_2Polypropylene (derived from H_2C=$CHCH_3$), polystyrene (derived from H_2C=CHC_6H_5), and polyvinyl chloride (derived from H_2C=CHClHCl) are examples of addition polymers. Starch and cellulose make up the second type of polymer. These are carbohydrate-related compounds with formulas that are multiples of CH_2O. The chemical formula for glucose is $C_6H_{12}O_6$. Starch and cellulose ($C_6H_{12}O_6$). Glucose molecules are linked together in both starch and cellulose, resulting in the loss of a water molecule at each linkage. [$C_6H_{12}O_5$] The synthetic material nylon is similar to this type of polymer. Water and a polymer (nylon-6,6).

Natural protein fibers with the repeating unit (-subunit-) include hair, wool, and silk, where r is a group of atoms attached to the repeating unit. Each bond or peptide linkage involves the joining of amino acids and the removal of a water molecule; for example, the tripeptide chain is made up of three units of the amino acid glycine (NH_2NH_2COOH n). In solvolysis and hydrolysis, the reactant also serves as the solvent in Solvolysis. When a solvent is present, the reaction is usually named after it; for example, hydrolysis is the reaction name when water is present. The hydrolysis reaction is represented by the reversible chemical reaction AB + HOH = AH +BOH if a compound is represented by the formula ab (where a and b are atoms or atom groups) and water is represented by the formula HOH. The reaction between water and esters is an example of organic compound hydrolysis. The general formula for esters is RCOOR ′, where R and R′ are combining groups. (such as CH3). The hydrolysis of an ester produces acid and alcohol.

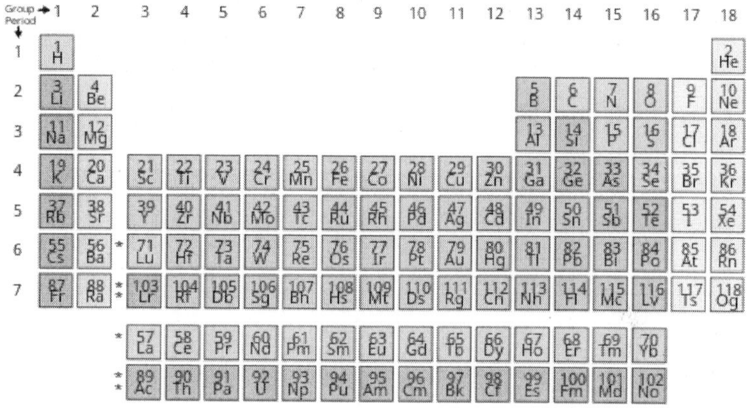

Figure 5.10: Chemical products on an element table: A periodic table colour-coded to show some commonly used sets of similar elements. The categories and their boundaries differ somewhat between sources.[88] Alkali metals Alkaline earth metals Lanthanides Actinides Transition metals Other metals Metalloids Other nonmetals Halogens Noble gases.

Source: By Offnfopt - Own work, Public Domain, https://commons.wikimedia.org/w/index.php?curid=62296883

All biological and chemical processes require hydrolysis reactions. Simple sugars are formed by the breakdown of complex sugars and starches. These processes are typically accelerated by the presence of enzymes or biological catalysts. Hydrolysis reactions influence acid-base behavior. Weak acid anions dissolve in water to form basic solutions in the same way that acetate doesDespite being a reactant-favored reaction, it occurs until a solution containing acetate exhibits fundamental properties (such as turning red litmus paper blue). A variety of fundamental properties of common substances can be explained by hydrolysis reactions (Metz et al., 2014).

Salts of the ions borate, phosphate, and carbonate, for example, produce basic cleaning solutions that have been used for centuries. Tartrate and citrate ions can be found in a variety of foods.

Classification based on the reaction mechanism. Reaction mechanisms describe the reassembly and rearrangement of atoms during the formation of products from reactants. The names chain reactions and photolysis come from the mechanisms that drive them. Processes that occur sequentially: Each product of a chain reaction serves as a reagent in the subsequent reaction. There are three distinct processes involved in chain reactions: an initiation step, a series of chain-propagation steps, and a termination step. Teflon is created through the chain reaction of polymerizing tetrafluoroethylene. Peroxide (a compound composed of two oxygen atoms joined together by a single covalent bond. Peroxides start the reaction by producing extremely reactive free radicals (species with an unpaired electron). There are numerous methods for breaking the chain.

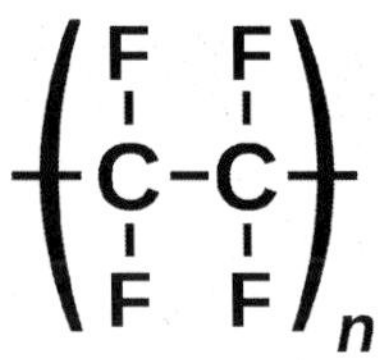

Figure 5.11: Teflon.

Source: By Vectorization: OmenBreeze - Own work based on: Teflon structure. PNG by Dubaj~commonswiki, Public Domain, https://commons.wikimedia.org/w/index.php?curid=84729767

5.4 PHOTOLYSIS REACTIONS

Absorption of electromagnetic radiation initiates and sustains photolysis reactions. In the section on kinetic considerations, the atmospheric decomposition of ozone to oxygen is used as an example. Another example is the photosynthesis of chloromethane from methane and chlorine. Is $CH_4(g) + Cl_2(g)$ -->$CH_3Cl(g)$ + HCl(g),HCl. This is an unintentional chain reaction. When a chlorine molecule (Cl_2) is exposed to ultraviolet radiation, it undergoes an endothermic reaction that produces chlorine atoms. After formation, not all chlorine atoms recombine to form chlorine molecules. When a chlorine atom collides with a methane molecule, chain propagation occurs in two stages (Bennett et al. 2010). The methyl radical (CH_3) is produced during the initial phase of propagation (. This free-radical species reacts with a chlorine molecule to produce the product and a chlorine atom, allowing the chain reaction to continue in multiple steps. The combination of two methyl radicals to form ethane (CH_3CH_3) and the combination of methyl and chlorine radicals to form chloromethane are two possible termination steps.

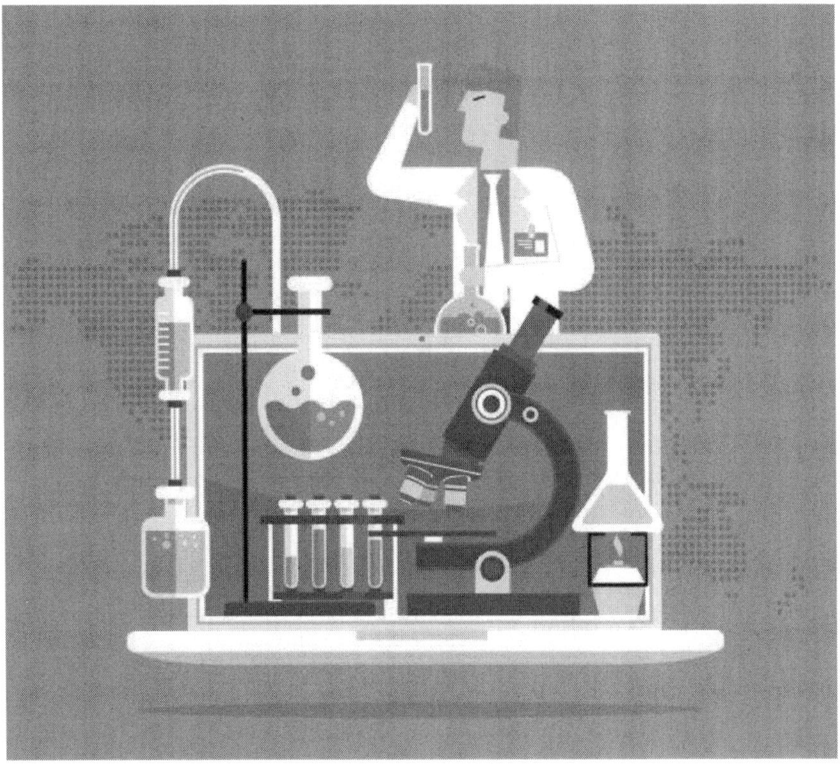

Figure 5.12: Chemical analysis by a laboratory technician.

Source: https://media.istockphoto.com/vectors/medical-laboratory-vector-id1395547195?k=20&m=1395547195&s=612x612&w=0&h=BWeXW4qKAZgrdaAvL896Z4tK_kTI9e6MO_vvXZ9FiQY=

CHAPTER 6

ELECTRONIC CONFIGURATION AND PERIODIC TABLE

CONTENTS

6.1 ELECTRON CONFIGURATION

Electron configuration is defined in quantum chemistry and atomic physics as the distribution of electrons of an electron of an atom or molecule with regards to molecules or atomic orbitals. For instance, the electronic configuration of neon is $1s^2\ 2s^2\ 2p^6$. This means that the 1s, 2s and 2p subshells have been occupied by 2, 2 and 6 electrons respectively. When describing electrons, the electronic configuration can define it as moving independently in an orbital in an average field created by other orbitals. Configurations can be expressed mathematically by Slater determinants or configuration state functions. Laws of quantum mechanics describe systems with only one electron and in certain conditions, the electrons are able to move from one configuration to another through the absorption or emission of quantum energy in the form of a photon. Usually, a system with one electron is associated its electron configuration. Knowledge on electron configuration is very important for different fields such as chemistry and physics (Congrains et al., 2013). It enables one properly to understand the structure of the periodic table of elements. Knowledge of electron configuration is also useful in describing the chemical bonds holding atoms together. When applied to bulk materials, knowledge of electron configuration is used in explaining the peculiar properties of semiconductors and lasers.

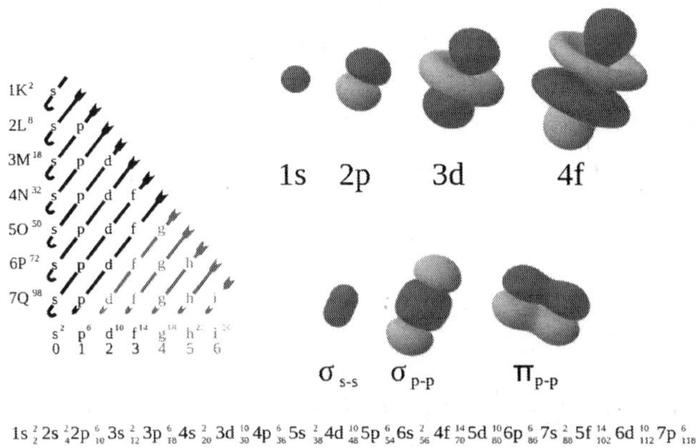

Figure 6.1: The electron configuration explains the distribution of electrons in an atom or molecule.

Source: By Patricia.fidi - This W3C-unspecified vector image was created with Inkscape., Public Domain, https://commons.wikimedia.org/w/index.php?curid=1239876

The concept of electron configuration was introduced in the Bohr model of the atom. It makes use of terms such as shells and subshells. Advances have been made in understanding the quantum-mechanical nature of electrons that have been used in further explaining the electron configuration of atoms. The term electron shell as used in electron configuration is used to refer to the set of allowed states that share the same principal quantum number that can be occupied by an electron. The quantum number usually denoted by the letter n is used to refer to the number before the letter in the orbital description. Usually, an atom's nth electron shell is able to accommodate $2n^2$ electrons. For example, the first shell can occupy 2 electrons, the second shell can 8 electrons while the third shell carries 18 electrons. It is noted that there is a factor of two that arises due to the fact that the allowed states are double (Mendoza & Núñez, 2009). This is attributed to the electron spin- each atomic orbital admits up to two, otherwise identical electrons with opposite spin. One of the electrons spins with +1/2 denoted by an up-arrow while the other spins with -1/2 denoted with a down-arrow.

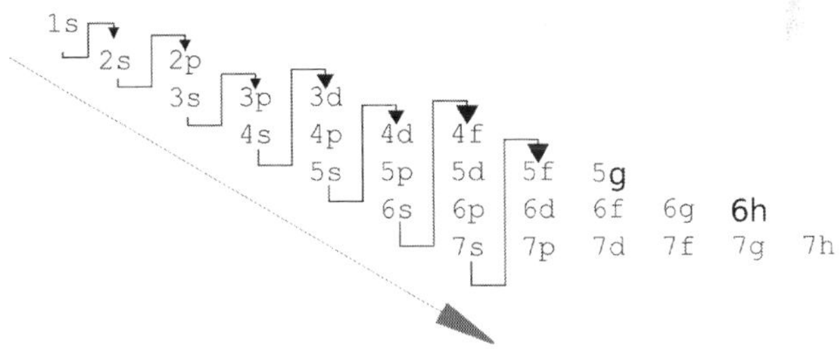

Figure 6.2: The electron configuration makes use of shell and subshells in the distribution of electrons.

Source: By Sharayanan - Own work, CC BY-SA 3.0, https://commons.wikimedia.org/w/index.php?curid=2022314

Another term used in defining an electron configuration is the subshell. The subshell is defined as the set of states defined by a common azimuthal quantum number, I, within the shell. Values of I can range from 0to -n-1. This means that values can be I=0, 1, 2, 3 and so on corresponding to the s, p, d and f descriptions respectively. For example, the 3d subshell could have n= 3 and I= 2. However, there is a limit with regards to the number of electrons that can be placed in a subshell. The maximum number of electrons can be calculated using the formula 2(2I +2). This means that in an s subshell the maximum number is electrons if two while in a p subshell the maximum number of electrons is six, ten electrons in a d subshell and in an f subshell the maximum number of electrons will be fourteen.

The equations of quantum mechanics are used to generate the number of electrons that can occupy each shell and each subshell. One of the equations is the Pauli Exclusion Principle that states that no two electrons in the same atom can have the same values of the four quantum numbers.

6.2 ENERGY OF GROUND STATE AND EXCITED STATES

When studying the electron configurations of atoms, we note that energy is associated with an electron is usually associated with the orbital it is in. The sum of the energy of each electron while neglecting the electron-electron interaction is used in approximating the energy of a configuration. The term ground state is used to refer to the configuration that corresponds to the lowest electronic-electron interactions. The other configurations are referred to as the excited state. For example, the ground state configuration of an atom of sodium is given by $1s^2\ 2s^2\ 2p^6\ 3s^1$ that can be deduced from the Aufbau principle. The first excited state is obtained through the promotion of a 3s electron to the 3p orbital to generate a configuration which is $1s^2\ 2s^2\ 2p^6\ 3p^1$ and abbreviated as the 3p level.

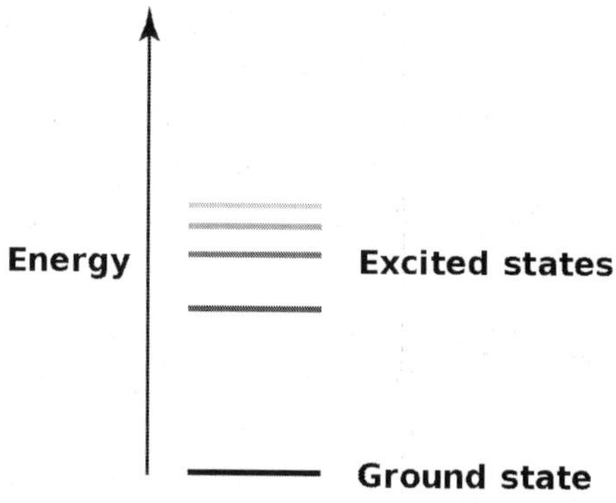

Figure 6.3: Electrons at the ground state are those that exist in low energy levels.

Source: By SVG: RehuaOriginal: Rozzychan - This file was derived from: Energylevels.png, CC BY-SA 3.0, https://commons.wikimedia.org/w/index.php?curid=35220791

An atom can move from one configuration to another by emitting or absorbing energy. This can be explained by the sodium-vapor lamp where the atoms of sodium are excited to the 3p level through an electrical discharge and afterwards return to the ground state through the emission of yellow light with a wavelength of 589nm. The excitation of valence electrons usually involves energies corresponding to photons of visible or ultraviolet light (Mansour et al., 2010). There are cases where the core electrons are excited though such a process requires much higher energies that tend to correspond to x-ray photons. Such a case involves the excitation of sodium to the 3s level and from the excited $1s^2\ 2s^2\ 2p^5\ 3s^2$ configuration. Ground state configuration is often referred to as the configuration of a molecule or an atom.

6.3 OPEN AND CLOSED SHELL

In the electronic configuration discussion in the field of physics, the terms open shells and closed shells are often used. When referring to an atomic orbit, the term open shell is used to refer to a valence shell that is not completely filled with electrons. It is also used to refer to a shell that has not given all of its valence electrons when it forms chemical bonds with other atoms or molecules during a chemical reaction. This means that the term closed shell is used to refer to a completely filled valence shell. The configuration of a closed shell is usually very stable. In the case of molecules, the term open shell is used to refer to unpaired electrons. This is well explained in the molecular orbital theory that leads to molecular orbital being singly occupied. The manner in which open shells are managed in molecular orbital theory is very different from the manner in which it is handled in computational chemistry where it is either handled by the restricted open-shell Hartree-Fock method or the unrestricted Hartree-Fock method. On the other hand, the closed-shell configuration corresponds to a state where all the molecular orbitals are empty or doubly occupied. Comparison between open shell and closed shell molecules show that open shell molecules are usually difficult to study computationally.

6.4 NOBLE GAS CONFIGURATION

The electronic configuration of noble gas is known as the noble gas configuration. A common characteristic of chemical reactions is that there is a tendency of chemical elements to acquire stability. A number of atoms obey the octet rule while those classified as transition metals obey the 18-electron rule. On the other hand, noble gases are very different as they are less reactive compared to other elements. The noble gas configuration explains their less reactive nature. However, oganesson is said to be highly reactive as it possesses relativistic effects for heavy atoms. Systems are said to possess the tendency to acquire the state of minimum energy or a state of stability. For this reason, chemical elements take part in chemical reactions with the goal of acquiring stable electronic configuration that is similar to that of its nearest noble gas (Marian, 2012). An example of an atom with this behavior is the hydrogen atom. If two atoms of hydrogen are reacting with one oxygen atom, the reaction leads to the formation of a water molecule. A neutral atomic hydrogen had one electron in the valence shell. For water to be formed, sharing is done between the second electron coming from oxygen making its configuration similar to that of its nearest noble gas helium having two electrons in the valence shell. In the same manner, a neutral atom of oxygen has 6 electrons in the valence shell and gets two electrons from the two hydrogen atoms. This makes its configuration similar to that of its nearest noble gas neon that has 8 electrons in the valence shell.

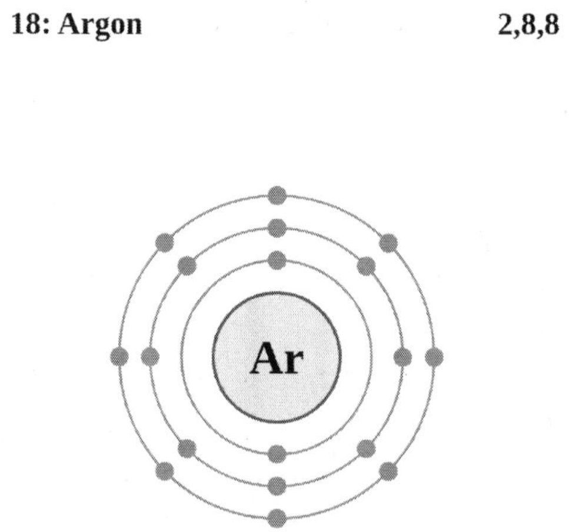

Figure 6.4: The noble gas electron configuration gives the example of a stable electron configuration.

Source: By Pumbaa (original work by Greg Robson) - Application: (generated by script), CC BY-SA 2.0 uk, https://commons.wikimedia.org/w/index.php?curid=717148

6.5 ELECTRON CONFIGURATION IN MOLECULES

The electronic configuration in molecules is very different from that of atoms. The electronic configuration of molecule is more complex than ordinary atom. This attributed to the fact that each molecule has different orbital structure. When descriptiondescriptionling molecules, their orbitals are descriptionled according to their symmetry rather than the conventional atomic orbitals descriptions used when descriptionled atoms and monatomic ions. For this reason, the electron configuration of a dioxygen molecule, O_2, can be written as $1\sigma g^2\ 1\sigma u^2\ 2\sigma g^2\ 2\sigma u^2\ 3\sigma g^2\ 1\pi u^4\ 1\pi g^2$ or $1\sigma g^2\ 1\sigma u^2\ 2\sigma g^2\ 2\sigma u^2\ 1\pi u^4\ 3\sigma g^2\ 1\pi g^2$. Two electrons in the two degenerate π*- orbitals is represented by the term 1πg2. From the Hund's rule, we learn that such electrons have parallel spins in the ground state. This means that dioxygen is paramagnetic meaning that it has a net magnetic moment. Major success in the molecular orbital theory was accredited to the explanation of the paramagnetism of dioxygen. It is worth noting that the electronic configuration of polyatomic molecules can change without the need for emission they absorb a photon through vibronic couplings.

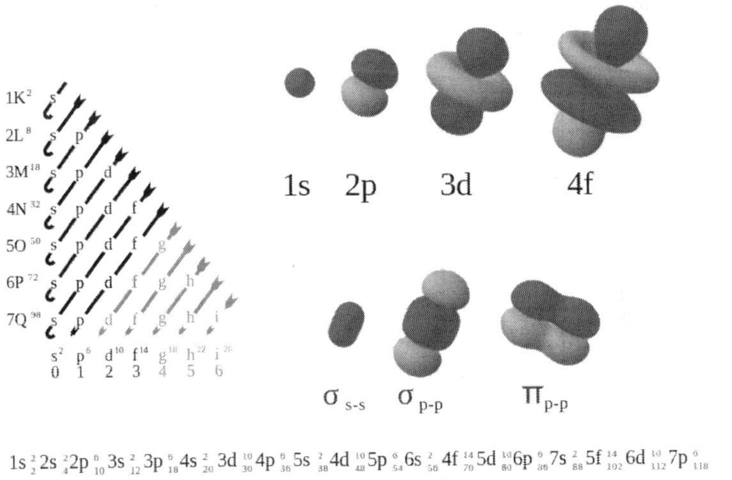

Figure 6.5: Molecules have a complex electron configuration.

Source: By Patricia.fidi - This W3C-unspecified vector image was created with Inkscape., Public Domain, https://commons.wikimedia.org/w/index.php?curid=1239876

6.5.1 Electron Configuration in Solids

Solids are not as complex as molecules. Instead, the electron states become very numerous. This means that they are no longer discrete. They are able to effectively blend to continuous ranges of possible states also known as an electron band. This means that the notion of electron configuration ceases to be relevant creating need for the band theory.

6.5.2 Application of Electron Configuration

There are various applications of electronic configuration mostly with rationalization if chemical properties in organic and inorganic chemistry. When the electron configuration is used alongside some simplified form of molecular orbital theory form part of the modern equivalent to the valence concept used in describing the type and number of chemical bonds that an atom can be expected to form. This approach has also been used in the field of computational chemistry that attempts to make quantitative estimates of chemical properties (Campos & Vasconcelos, 2010). It is a deviation from the normal calculations that heavily depended on the linear combination of atomic orbitals approximation. This made use of an ever-larger and more complex basis set of atomic orbital that can be used as the starting point. The final steps of the calculations involve the assignment of electrons among the molecular orbitals following the Aufbau principle.

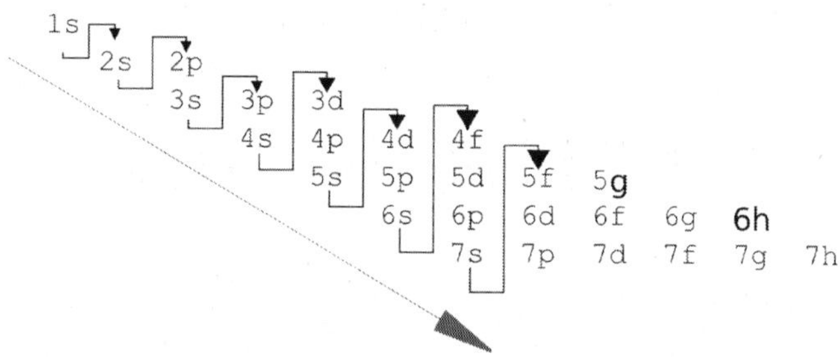

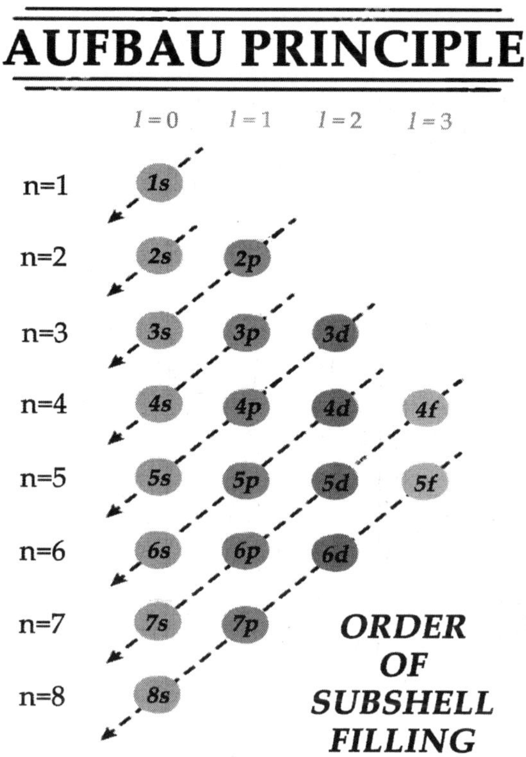

Figure 6.6: The electron configuration is used in solving Chemistry calculations.

Source: By Sharayanan - Own work, CC BY-SA 3.0, https://commons.wikimedia.org/w/index.php?curid=2022314

It is important to note that not all methods in calculational chemistry make use of electron configuration. A good example is the density functional theory, one of the methods that discards the model. Molecules or atoms with more than one electron usually have the motion of their electrons being correlated and such a picture is no longer exact. In such a case, a very large number of electronic configuration is needed to exactly describe a multi-electron system. In such a system, no energy can be associated with a single configuration. In most cases, the electronic wave function is dominated by a very small number of configurations. For this reason, the notion of electronic configuration is considered very essential for multi electron systems.

One of the areas of application of electron configuration is the interpretation of the atomic spectra. In such an application, it is very important to supplement the electron configuration with one or more symbols all of which are needed in describing the different energy levels available in an atom. For any electron configuration, term symbols can be calculated (Colussi et al., 2013). It is important to note that not all energy levels are observed in practice. The calculation is through the analysis of atomic spectra that the ground state electron configurations of the elements are determined experimentally.

6.5.3 Writing an Electron Configuration

When writing the electron configuration of an element, the symbols used in writing the electron configuration start with the shell number followed by the type of orbit and the superscript indicates the number of electrons in the orbital. When writing the electron configuration of an element, the electron configuration symbol is of great importance. Usually, the periodic table plays a vital role in writing the electron configuration of an atom. For example, in the periodic table, oxygen is seen to have 8 electrons. For the 8 electrons, we can say that oxygen has the order of 1s, 2s and 2p making oxygen's electron configuration to be $1s^2 2s^2 2p^4$.

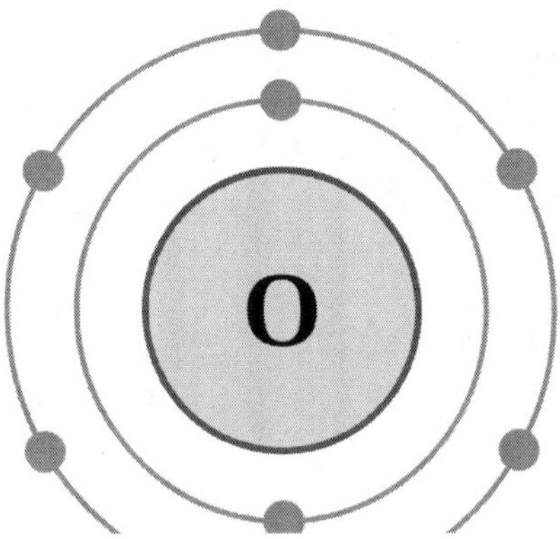

Figure 6.7: The electron configuration symbol is used in writing the electron configuration of an atom.

Source: https://commons.wikimedia.org/wiki/File:Electron_shell_008_Oxygen_-_no_label.svg

When writing the electron configuration of an element, it is important to note that there are special cases. This is usually the case when dealing with ions. Special cases of electron configuration are noted when demonstrating the reason for the formation of certain ions in the first place. When writing the full electron configuration of an anion, one will be adding additional electrons hence the configuration is continued. This can be demonstrated by oxygen known to form 2- when making an ion. This means that 2 electrons will be added to its normal configuration. The new configuration becomes O_2-$1s^2 2s^2 2p^6$. When oxygen has 10 electrons, its electron configuration becomes similar to that of Neon. Usually ions are formed by attaining stability through the loss or gain of electrons becoming like noble gases.

For the case of cations, the electron configuration can be made in the basis of the number of electrons though there is a deviation in the manner of configuration. The process begins the writing the normal electron configuration and after the electrons are removed, they are taken from the outermost shell. This is not the same way they are added. For example iron has 26 electrons and its normal electron configuration would be Fe $1s^2 2s^2 2p^6 3s^2 3p^6 4s^2 3d^6$. However, when 3+ iron is made for iron then electrons would be taken from the outermost shell (Chen et al., 2015). This means that electrons would be taken from the 4s shell and not the 3d shell. This leads to the electron configuration being Fe^{3+} $1s^2 2s^2 2p^6 3s^2 3p^6 3d^5$.

There are some short cuts to writing electron configurations of elements. For instance, when writing the configuration of elements of the lower table configurations, the total configuration would be fairly long. In such a case, the noble gas configuration can be used in abbreviating the configuration. One will just have to finish the configuration from where the noble gas leaves it. With noble gas configurations, it is important to note that there are some exceptions when abbreviating the configurations of noble gases. The exceptions are usually with regards to the order of fill. A good example is in block d where are some exceptions in the manner in its filling for groups containing Copper and Chromium

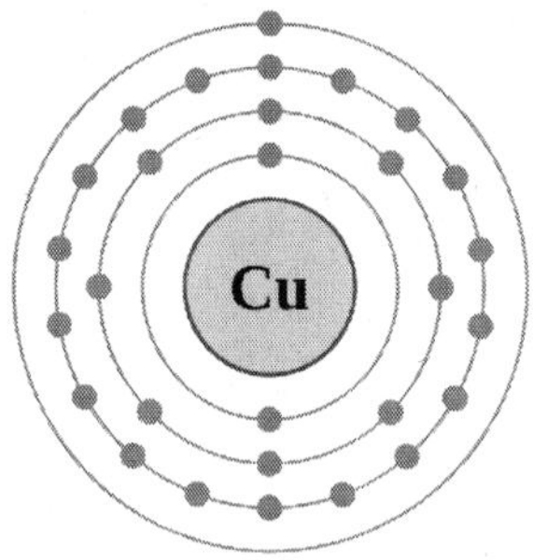

Figure 6.8: There are exceptions in the filling of copper when writing electron configuration.

Source: https://commons.wikimedia.org/wiki/File:Electron_shell_029_Copper_-_no_label.svg

The order of filling an atom can get well represented by orbital diagram. Orbital diagrams are also known as the little boxes. The boxes are used when representing the orbit. They are used to show the placement of electrons in them. Usually, the order of filling is usually the same. However, with electrons, they are placed singly into the boxes before they are filled with electrons. This is referred to as the Hund's Rule that states, "Half fill before you full fill". This rule was established on the basis of energy calculations that indicated that it was the manner in which atoms actually distribute their electrons in the orbitals.

6.6 THE PERIODIC TABLE

When writing the electron configuration of atoms, the periodic table is very useful. The periodic table is also referred to as the periodic table of chemical elements. It is a tabular display of chemical elements. The periodic table is widely used in physics, chemistry and other sciences. The periodic table is generally seen as an icon of chemistry. It can also be considered a graphic formulation of the periodic law that states that the properties of the chemical elements tend to exhibit a periodic dependence on their atomic numbers. The periodic table is divided into four rectangular areas known as blocks. The table has rows also known as periods while the columns are called groups (Chaplin, 2006). Elements classified to bring the same column group of the periodic table will exhibit the same chemical characteristics. There are trends throughout the periodic table in that non-metallic characters that keep their own electrons will increase from the left to the right across a period and it runs from down to up across a group. On the other hand, those in the metallic character known to surrender electrons to other atoms will increase in the opposite direction. The electronic configuration of atoms will explain these trends.

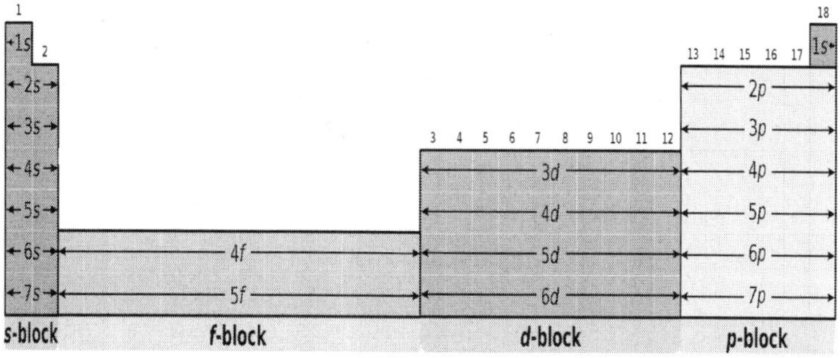

Figure 6.9: The relationship between electron configuration and periodic table enables proper configuration of elements.

Source: By User:DePiep - https://commons.wikimedia.org/wiki/File:Periodic_Table_2.svg, CC BY-SA 3.0, https://commons.wikimedia.org/w/index.php?curid=33928254

The periodic table was developed and was generally accepted was that of the Russian chemist Dmitri Mendeleev. It was developed in 1868. This periodic table was developed alongside the periodic law as a dependence of chemical properties on atomic mass. During that time, not all elements were known explaining the gaps in his periodic table. He used the periodic law in predicting the properties of some of the missing elements. The developed periodic law was recognized to be a fundamental discovery in the late 19th century. Its discovery was explained by the discovery of the atomic number. Its importance was also highlighted when pioneering work in quantum mechanics of the early 20th century that resulted in the release of information on the internal structure of the atom (Matranga & Corsi, 2012). A discovery was made in 1945by Glenn T. Seaborg that showed that the actinides were in fact f-block rather than d-block elements. This was the recognized form of the periodic table developed. Both the periodic law and periodic table are central and considered an indispensable part of modern chemistry.

With the continuous progress in science, the periodic table continues to evolve. Elements with atomic number 94 is said to exist naturally. Some elements are synthesized in the laboratory. Currently, all the first 118 elements are known this means that the first seven rows of the table are complete. However, chemical characterization is needed for the heaviest elements. It is used in confirming that their properties match their positions. There is a likelihood that the table will stretch beyond the seven rows. There are other alternative representations of the periodic law in existence.

When discussing the periodic properties of the periodic table, we note that there is a relationship between the periodic table and electron configuration. One can say that the periodic table was made to allow elements with similar electron configurations to be aligned into the same group.

6.7 ELECTRON CONFIGURATION PERIODIC TABLE

The periodic table shows the last orbital filled for each element. It is also used to demonstrate how the configuration of each element has been aligned such that the last filled orbit is the same except for the shell. This is done so that the configuration of an element gives the element its properties and similar properties brought about by similar configurations. There are periodic properties influenced directly to the electron configuration.

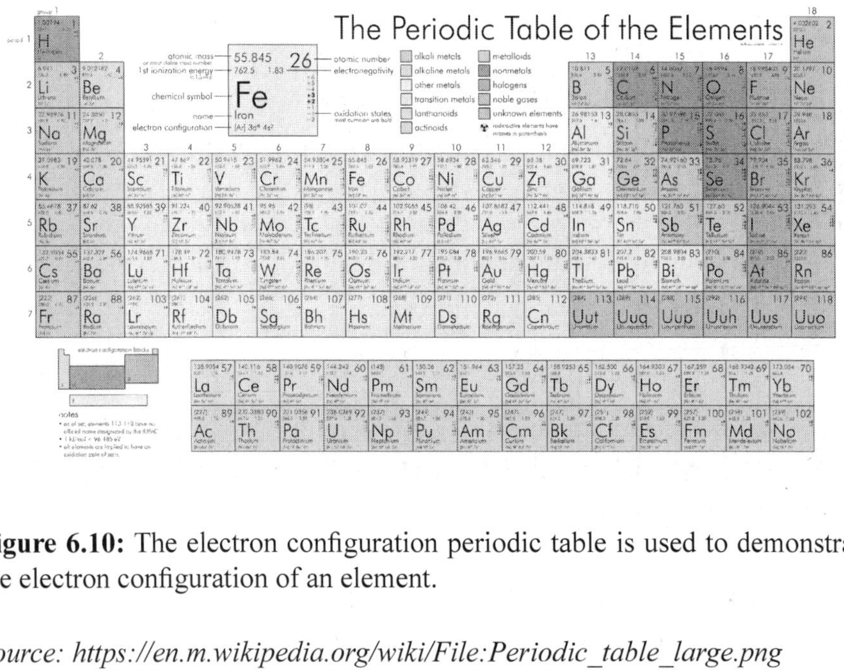

Figure 6.10: The electron configuration periodic table is used to demonstrate the electron configuration of an element.

Source: https://en.m.wikipedia.org/wiki/File:Periodic_table_large.png

The electron configuration of elements is used in placing them in the periodic table. When placed there, they exhibit periodic recurrences useful in explaining the trends of properties across the periodic table. An electron can be said to inhibit an atomic orbital that characterized the probability that it can be found in any particular region of the atom. The energies of electrons are quantized. This means that they can only take discrete values. Electrons obey the Pauli Exclusion Principle that states that different electrons can take in various energy levels known as shells. The shells are further subdivided into individual subshells. Each of the subshells contains a certain kind of orbit. Each orbital has the ability to carry up to two electrons. The orbitals can also be distinguished by a quantity know. As spin which can be up or down (Lovley, 2003). While in the atom, electrons arrange themselves in such a way that the total energy they have is minimized. This means that they occupy the lowest-energy orbitals available unless it is subjected to energy. The valence electrons also known as the outermost electrons have the need to break free of the nucleus and participate in chemical reactions with other atoms. There are electrons known as the core electrons.

Currently, the known elements have seven shells occupied. The first shell usually carries one orbital which is the spherical s orbital. The first shell is also known as the 1s orbital and has the ability to hold up to two electrons. The second shell also contains a 2sorbital. However,1s carries three dumbbell-shapes p orbitals. This means that it can fill up to eight electrons. The third shell carries one 3s orbital, three ,2/3, p orbitals and five 3d orbitals. This means that 3s can carry 2*2 + 2*3 + 2*5 = 18. The fourth shell carries one 4 orbital, three 4p orbitals, five 4d orbitals and seven 4f orbitals. This means that it has a capacity of 2*1 + 2*3 + 2*5 2*7 = 32. As the number of shells increases, there are more types of orbitals that carry on with the patterns. It is important to note that for the element with increased shells, their orbitals are not filled in the known elements. Quantum numbers are used to describe the subshells. Four numbers are used in completely describing an electron in an atom. The principal quantum number n is the shell, the other orbital type is the azimuthal quantum number, the magnetic quantum number describes the kind of orbital the electron is in and finally there is the spin quantum number s.

6.8 PERIODIC TRENDS

Valence electrons of elements are involved in chemical reactions. Elements with similar outer electron configurations are expected to react in the same manner while the compounds have similar proportions of elements in them. Such kind of elements are usually placed in the sane group. This brings about the clear similarities and trends in chemical behavior as one looks down the group. The properties of elements are said to exhibit periodic recurrences as the analogous configurations are noted to return at regular intervals. This brought about the name periodic law and the periodic table. The periodic recurrences were noted to occur even before the theory explaining them was developed. A good example is the valence of an element. The valence can be defined as the number of hydrogen atoms combining to form a simple binary hydride (Lischka et al., 2011). It can also be defined as twice the number of oxygen atoms that can combine with it resulting in the formation of a simple binary oxide. In most cases, the valences of main-group elements are usually directly related to the group number. For example, hydrides in the main-groups 1-2 and 13-17 follow the formulae MH, MH_2, MH_3, MH_4, MH_3, MH_2, and MH. There is an increase in valence in the highest oxides. Such oxides have the formulae M_2O, MO, M_2O_3, MO_2, M_2O_5, MO_3, M_2O_7.

The electron configuration is used in giving a possible explanation from the number of electrons available and can be used in bonding. When getting full explanations, one should consider the energy that would be released when compounds are formed from elements having different valence electrons. This information is very useful in addition to using the electron configuration. Currently, the notion of valence has been extended by that of the oxidation state. This is the formal charge left on an element when other elements in a compound are released as their ions. Elements that share the same group may have the same valence configuration. This means that they have similar chemical behavior. A good example is with the alkali metals in the first group that have one valence electron. They are also known to form a homogeneous class of elements.

The alkali metal are known to be soft and reactive metals. It is important to note that there are several factors involved. Groups of elements can also be heterogeneous. An example is with stable elements of group 14 that comprises of a nonmetal(carbon), two semiconductors (silicon and germanium) and finally two tin metals such as lead and tin. These heterogeneous elements are united by having four valence electrons. With regards to electron affinity, there is a notable trend in the periodic table. Electron affinity is the opposite of ionization energy. It refers to the release of energy when adding an electron to the atom (Lisio et al., 2019). When an electron passes by an atom, the passing electron will be readily attracted to an atom. This happens because the electron feels the pull of the nucleus that is stronger more so when there is an available partially filled outer orbital where it can suitably fit. This means that from down to up the periodic table and left to right, there is an increase in electron affinity. However, there is an exception in the last column characterized by Nobel gases that have full shell and have no space to allow the entry of an electron. This means that the halogens in the next-to-last column have the highest electron affinities.

The electron configuration of an element can help identify its electron affinity. For atoms such as the noble gases, they have no electron affinity. This means that they cannot for stable gas-phase anions. Noble gases with high ionization energies and no electron affinity tend to have a little inclination towards losing or gaining electrons. They are said to be generally unreactive. There are exceptions in the trend of electron affinity in the periodic table. For instance, fluorine and oxygen have lower electron affinities than their heavier homologues chlorine and sulfur. This is attributed to the fact that they are small atoms explaining the newly added electron that would experience significant repulsion from already present ones (Laganà et al., 2019). In the case of non-metallic elements, their electron affinity somewhat correlated with reactivity. The correlation is not perfect due to other factors involved. A good example is with fluorine has lower electron affinity compared to chlorine but has higher reactivity.

CHAPTER 7

COVALENT BONDING

CONTENTS

In a covalent bond, each atom contributes an equal number of electrons to the total link. The pair of electrons involved in this sort of covalent bonding is known as a shared pair or bonding pair. Covalent bonds are also known as molecular bonds. The outer shell of the atoms' stability is obtained through sharing bonding pairs, as in noble gases.

Elements with high ionization energies cannot transport electrons, whereas elements with low electron affinity cannot receive electrons. Exchanging electrons with atoms of different elements or the same element results in an octahedral arrangement of electrons in the valence shell, resulting in a stable octet configuration. Two molecules must exchange electron pairs to form covalent bonds (Kleber & Johnson, 2010).

Covalent bonds are those in which atoms exchange electron pairs. The binding is caused by the nuclei's electrostatic attraction to the electrons. It is in charge of the creation of atomic bonds. According to the octet rule, atoms must exchange electrons to maintain a stable electronic state. Covalently bonded molecules form covalent compounds. In organic chemistry, covalent bonds are more prevalent than ionic ones. This process is also known as molecular bonding.

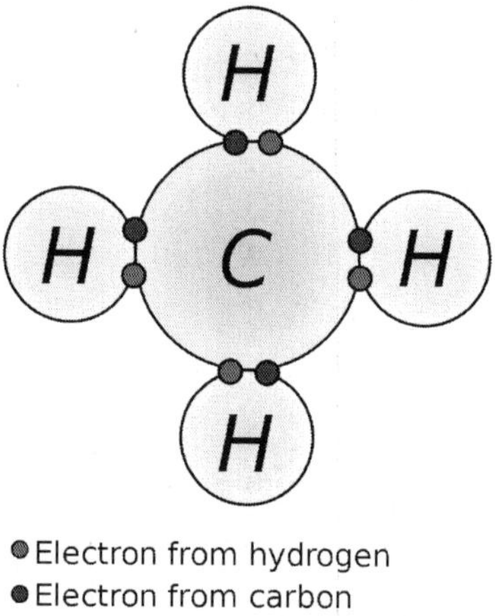

Figure 7.1: Structure of covalent bonds.

Source: By DynaBlast - Created with Inkscape, CC BY-SA 2.5, https://commons.wikimedia.org/w/index.php?curid=995735

7.1 DO YOU UNDERSTAND WHY COVALENT BONDS FORM?

When the difference in electronegativity between two atoms is insufficient for electron transit, a covalent bond is created.

Electronegativity is a measure of an atom's capacity to attract electrons. Atoms create covalent bonds to strengthen their stability by forming a whole electron shell by sharing their outermost (valence) electrons.The attraction between positively charged nuclei and shared electrons is stronger than the repulsion between them in covalent bonds. This attraction aids in the retention of the molecules. The amount of energy required to break the connection, or the amount of energy required to separate the connected atoms determines the strength of a covalent bond.

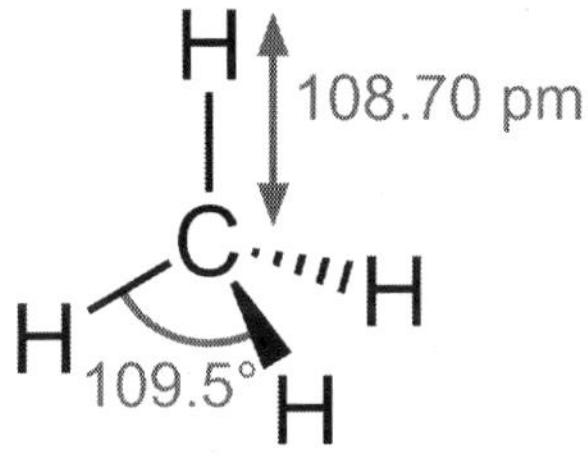

Figure 7.2: Structure of the covalent bonding of methane.

Source: By Original: Benjah-bmm27 Vector: Jynto - Own work based on: File:Methane-CRC-MW-dimensions-2D.png, Public Domain, https://commons.wikimedia.org/w/index.php?curid=12422898

7.2 COVALENT BOND CHARACTERISTICS

In the structure of stable covalent compounds, covalent bonding is typically the controlling factor. A covalent bond has the following characteristics and qualities:

- It is possible to make multiple covalent bonds between two atoms when composed of two or more nonmetals, or one nonmetal and a metalloid. Breaking bonds requires a lot of energy, which is why it is so tough (Krylov & Gill, 2013).

Directional or Isomerism is the phenomenon in which one chemical formula may represent several distinct molecules.

- Covalent compounds have low melting and boiling temperatures, do not conduct electricity, and are insoluble in polar solvents such as water. Covalent substances can be dissolved using nonpolar solvents such as benzene and toluene.
- The reactions of covalent compounds are slow.

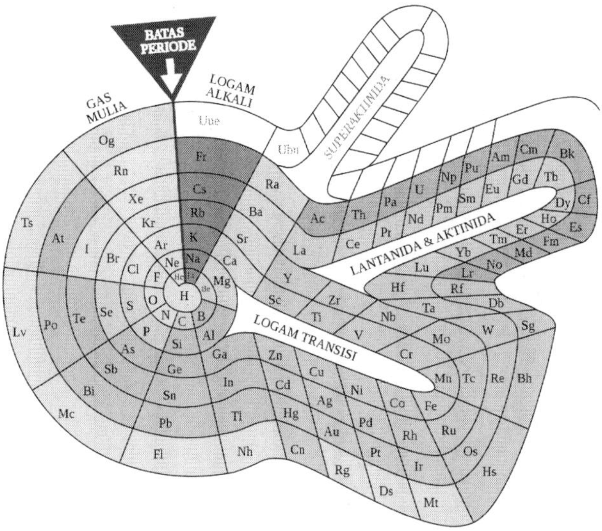

Figure 7.3: The elements table: Otto Theodor Benfey's spiral periodic table (1964).

Source: By DePiep - Own work, CC BY-SA 3.0, https://commons.wikimedia.org/w/index.php?curid=27766488

Lewis notation, commonly known as electron dot notation or Lewis dot structure for valence electrons, has dots around the atomic symbols (the ones in the outer shell). Atoms must share electrons to create covalent bonds. Multiple bonds are represented by various pairings, such as double bonds and triple bonds. This example figure does not include other representations of electron pairs that form bonds, such as solid lines. Lewis postulated that if an atom forms a sufficient number of covalent bonds, it can build a closed outer electron shell. In this Methane diagram, for example, the carbon is represented by an octet rule, which means it has four of its electrons and four from the hydrogens bonded to it (Chemat et al., 2012). The number of electrons around each hydrogen atom must be equal to the number of electrons surrounding each carbon atom (duet rule). The number of electrons in a complete shell, for example, is defined by the number of electrons in an atom's outer (and sole) shell. For example, a carbon atom's n = 2 shell has eight electrons, but the n = 1 shell contains two electrons. Only quantum mechanics can anticipate the shapes and characteristics of small molecules, which is required to fully comprehend covalent bonds. Because of the existence of shared electron pairs, covalent bonding may be better understood in terms of its qualitative features. In 1927, he offered a mechanical explanation for a chemical link (molecular hydrogen) for the first time. (Hydrogen molecule)

Their discoveries were founded on the valence bond principle, which states that a chemical bond is created when the atomic orbitals of the atoms involved overlap (Dubernet et al., 2010). In species with an odd number of electrons, one or three electron bonds can be inserted. The dihydrogen cation, H^+, has the greatest single-electron bonds. Because one-electron bonds have half the bond energy of two-electron bonds, they are also referred to as "half bonds."

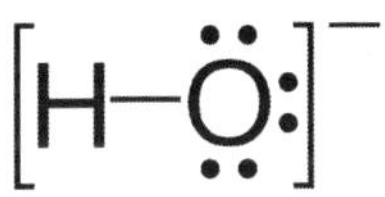

Figure 7.4: The covalent bond on hydroxide.

Source: By DoSiDo - Own work, CC BY-SA 3.0, https://commons.wikimedia.org/w/index.php?curid=2803509

Dilithium, for example, has a greater affinity for lithium than most other lithium compounds. This odd event can be attributed to hybridization and inner-shell effects. The helium dimer anion, $He^{+ 2}$, exhibits the most basic three-electron bonding. When two electrons form a "half bond," the third electron is in an anti-bonding orbital, which cuts the electron's link to the other two electrons in half. Nitric oxide, or NO, is another molecule with a 3-electron link in addition to two 2-electron bonds. The Para magnetism and formal bond order of O_2 are explained by two three-electron bonds and one two-electron bond. Chlorine dioxide has three electron bonds with bromine dioxide and iodine dioxide, their heavier cousins (Kharat et al., 2017). Odd-electron bonds are highly reactive molecules. Only atoms with the same electronegative charge may establish these bonds.

Covalent bonds are classified based on the number of electrons transferred, the polarity of the bonds, and the coordination of atoms. Based on the number of electron pairs they share, covalent bonds are classified into three types: The most prevalent form of link is covalent bonding (Single) When two electrons are shared between atoms, a single covalent connection, or simply a single bond, is formed. H_2, Cl_2, Br_2, I_2, HCl, NH_3, CH_4, and C_2H_6 are a few examples. A covalent bond is formed between two atoms. A double bond is formed when two atoms share two pairs of electrons or four electrons. The gases that make up the Earth's noble include O_2, CO_2, SO_2, and C_2H_4. A triple bond is formed when two atoms share three pairs of electrons or six electrons. CN, C_2H_2, and N_2 are a few additional examples (Kawser Hossain et al., 2016). Covalent bonds can be divided into three separate types based on their polarity and atom coordination: Bonds formed by polar atoms and ions A covalent bond is more likely to be polar when the electronegativities of the atoms sharing the electrons differ greatly, such as by 0.1 to 2.

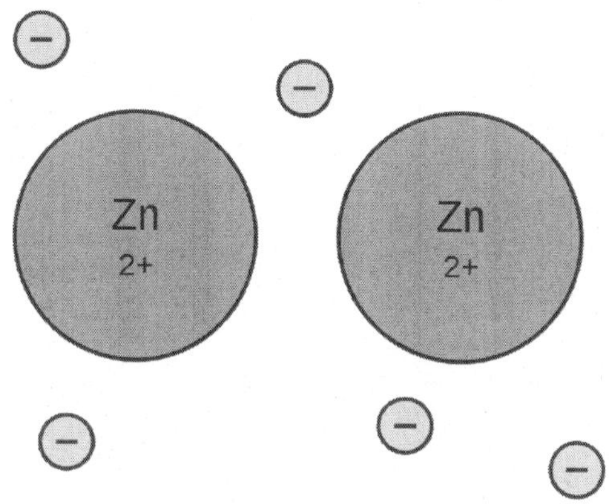

Figure 7.5: Metallic bond in zinc.

Source: By derivative work: Jcwf (talk)Metallic_bond_Cu.svg: King of Hearts - Metallic_bond_Cu.svg, Public Domain, https://commons.wikimedia.org/w/index.php?curid=4467807

When a bonded pair attracts a more electronegative atom, the more electronegative atom becomes slightly negative, while the other atom becomes slightly positive. Water, chloroform, hydrochloric acid chlorohydrate, HCl, and NH_3 are just a few examples. When there is no change in electron density, there is no difference in electronegativity between atoms. In this case, covalent bonds are nonpolar. H_2O, N_2, CO_2, and CH_4 are only a few of the numerous examples. In a Covalent Dative or Coordinate bond, one of the atoms provides the shared pair of electrons in this type of covalent bond.

A frequent example of this sort of interaction is ligand-ligand interactions. Covalent bonding can be accomplished in two ways: When two identical atoms share an electron, for example, consider the generation of different gases such as H_2, C_{12}, O_2, and others. The exchange of electrons between various kinds of atoms Consider the synthesis of CH_4, H_2O, and NH_3. Covalent bonding between carbon atoms Carbon must shed four electrons to produce C^{4-} because six protons cannot contain ten electrons (Dufrêne et al., 2017). Because it would take a lot of energy to lose four electrons to produce $C^{4+,}$ and because C^{4+} would only have two electrons bonded by the proton, it would be unstable again. Carbon must share one electron to reach its nearest noble gas state and form a covalent bond since it cannot receive or give electrons.

Characteristics of Covalent: Bonding

Atoms may share numerous electron pairs if sharing a single pair is insufficient to fulfill each atom's valence. Covalent bonds have the following properties: When two atoms make a covalent connection, they create an extremely strong chemical bond. At 80 kilocalories per mole of energy, covalent bonds transmit around 80 kcal/mol. Covalent bonds are directed because the bound atoms have unique relative locations. In terms of melting and boiling points, the majority of covalently bonded compounds have lower melting and boiling values. Covalently bonded substances do not conduct electricity because there are no free electrons available. Water does not dissolve covalent compounds. The Octet Rule's operation is unknown. Except for noble gases, every atom has less than eight electrons in its valence shell. This suggests that the valence shell configurations of the atoms are not long-term stable. As a result, when joined with other atoms, they may create stable electronic structures. To summarize: The octet rule states that an atom's valence shell should have as many electrons as possible, and that "the inclination of atoms from different elements to reach the stable configuration of eight electrons in their valence shells is the cause of chemical combination." Lewis utilized a basic set of symbols to represent an atom's outermost electrons. The chemical structure is known as the Lewis Dot Structure and Electron Dot Symbols.

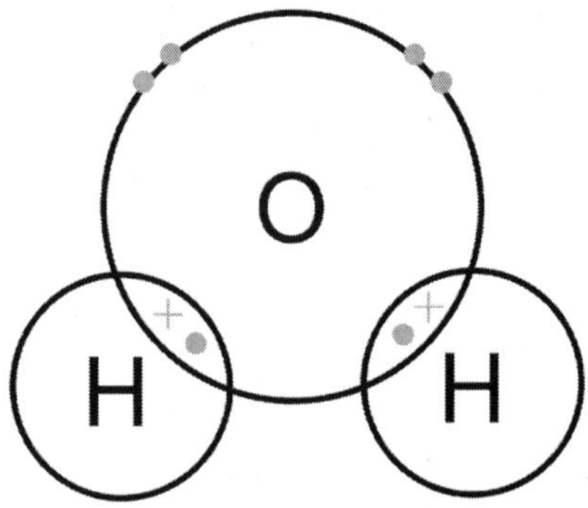

Figure 7.6: Lewis structure of a covalent bond.

Source: By Benjah-bmm27 - Own work, Public Domain, https://commons.wikimedia.org/w/index.php?curid=2112209

Covalent bonds are formed when two atoms share an electron pair in a Lewis dot structure. Two electrons are added to every single bond during the bond-formation process to establish the eight-atom configuration in the outermost shell of each element›s nucleus. Such electron dot structures are depicted using the octet rule. The valence shells of all atoms in the molecule, except for hydrogen, will possess eight electrons, according to this equation (Karsenti, 2008). To form the helium configuration, just two electrons are needed to complete the initial hydrogen shell. A stable octet would have one electron shared by elements in group 17 such as Cl. Two electrons would be shared by elements in group 16 such as O and S. For example, two oxygen atoms share two electrons with two hydrogen atoms to produce a water molecule, each of which has six electrons in its valence shell. They are available in a variety of sizes and shapes.

Covalent bonds can be divided into single, double, and triple covalent bonds based on the number of electron pairs shared by the bonds. A type of bond that has only one owner A solitary bond is formed when two atoms share only one pair of electrons. A single dash is a sign representing it (-). . This sort of covalent connection, although being less dense and weaker than double and triple bonds, is the most stable. So, for example, the HCl molecule contains one Hydrogen electron and seven Chlorine electrons on the side. Hydrogen and chlorine share an electron to form a single bond in this case. Two-bond configurations A double bond is formed when two atoms share two pairs of electrons. It's made up of two equal-sign dashes

(Kalra et al., 2016). Although double covalent bonds are more powerful, they are also less stable than single covalent bonds. Example: There are six valence electrons in one carbon atom, and four valence electrons in each of the two oxygen atoms. A carbon atom has four valence electrons, two of which are shared by two different oxygen atoms. Because each oxygen atom shares two electrons with carbon, CO_2 has two double bonds.

In oxygen molecules, each oxygen atom's valence shell has six electrons during the creation of an oxygen molecule. Each atom's octet requires an additional two electrons. To create the oxygen molecule, the atoms each share two electrons. Double bonds are formed when two atoms of oxygen share two electron pairs. Two hydrogen atoms exchange valence electrons with each carbon atom in the ethylene molecule, and the other carbon atom shares valence electrons with the other two carbon atoms. Because of this, there is a double bond between the carbon atoms. In a three-way tie, A triple bond is formed when two atoms share three pairs of electrons. In terms of stability, triple covalent bonds (shown by three dashes) are the weakest. To synthesize a nitrogen molecule, three electrons from each nitrogen atom with five valence electrons are given away, resulting in three electron pairs. A triple bond is formed as a result of the two nitrogen atoms. Bonds with polar covalent nature is formed when atoms with varying degrees of electronegativity unite and share electrons unequally, they create this type of covalent connection. Electrons will be more attracted to more electronegative atoms. Even if it's greater than 0, the electronegative difference between the atoms is less than 2.0. Thus, the shared electron pair of that atom will be in closer proximity (Huang & Zou, 2010).

Hydrogen bonds may occur as a result of an imbalance in the electrostatic potential. Fluorine, hydrogen, and oxygen all react with hydrogen atoms in this situation. Bonds that are nonpolar (Nonpolar) Covalent bonds are formed when two atoms have an equal number of electrons sharing one atom's valence shell. The electronegativity difference between any two atoms is a 0. This is the case when two diatomic elements (atoms with the same electron affinity) come together (diatomic elements), Nonpolar covalent bonds are found in gas molecules such as hydrogen and nitrogen.

The Polarization of Chemical Bonds: It is always the more electronegative atom that has the electron cloud closest to it in sigma bonds. An ever-present dipole results from the bond's being polarized. What Is the Difference Between Covalent and Ionic Bonds? Covalent and ionic bonds are two types of atomic bonding to consider. The architecture and behaviors of these bonds differ greatly. If the electron pairs of two atoms are connected in the same direction, covalent bonds are created. When two ions come into contact, they establish ionic bonds. It is the sharing of electron pairs between two non-metallic atoms, and additional covalent bonds with an electronegativity difference of more than 2.0, which constitutes covalent bonding (. When covalent bonds are formed, polyatomic ions are formed. Electrostatic interaction between ions with opposite charges creates an ionic bond. Nonmetals such as carbon, hydrogen, oxygen, and nitrogen create covalent bonds with one another and with other atoms in a molecule (Ebbesen, 2016). Following are examples of where covalent bonds can be established: One oxygen atom, two hydrogen atoms, and three nitrogen atoms Carbon is number four on the periodic table. Examples of covalent compounds are provided in the examples that follow. The chemical element hydrogen (H_2) is a gas.

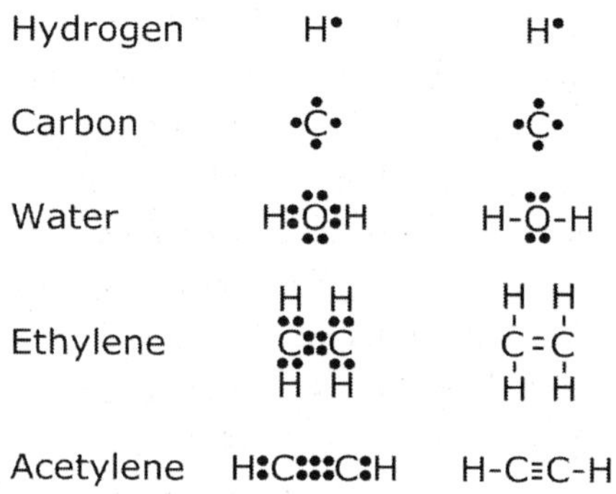

Figure 7.7: The types of covalent bonds.

Source: By DynaBlast - Own work, CC BY-SA 2.5, https://commons.wikimedia.org/w/index.php?curid=540013

Hydrogen (H_2) is the smallest and most abundant element. Helium, the nearest inert gas, has just one electron, thus it needs an extra electron to achieve the same electrical configuration. A molecule of hydrogen is formed when two hydrogen atoms join together to create a single bond. Oxygen is a chemical element (O_2) With an electron-requiring valence of two, the outermost oxygen (O) shell can only hold one pair of electrons at a time. To form a double bond, oxygen atoms combine and share their two valence electrons (N_2) To form an octet, two nitrogen atoms must be combined since each nitrogen atom has five valence electrons to work with. A molecule of nitrogen is the consequence of three electrons being shared between the two, resulting in three covalent bonds (Gentile et al., 2014). Three. Drinking water (H_2O) In a water molecule, there are two hydrogen atoms (H) and one oxygen atom (O). Unlike oxygen, hydrogen has only one electron in its orbital. A covalent bond is formed between oxygen and hydrogen because each hydrogen atom shares an electron with the other.

Consequently, there will be two one-sided bonds. There is a lot of CO_2 in the air (CO_2) There are two oxygen (O) atoms and one carbon (C) in CO_2. The valency of carbon is four, while the valence of oxygen is two. As a result, each oxygen shares two of its valence electrons with the carbon, forming a double bond. Every C=O bond is, thus, a double bond. Fourth is CH_4 methane. Methane is made up of one carbon (C) atom and four hydrogens (H) atoms. The valence of carbon is four, whereas the valence of hydrogen is one. Covalent bonds are formed when two atoms share the same number of electrons. Each of Methane's four covalent bonds will be made up of a single covalent bond. Sixth on the list of most abundant elements is ammonia (NH_3) To complete its valence shell, nitrogen (N) has five electrons in its outer orbital,but needs three more electrons. An additional electron (H) is shared with nitrogen, and three hydrogen atoms are needed to complete the outer shell. The sharing of electrons results in the formation of three single covalent bonds. Three covalent bonds unite carbon (C) and oxygen (O) to form CO. There are four electrons needed to complete the carbon's outermost shell because of its valence of four. With a valence of two, oxygen's outermost shell needs two electrons to be completed. As a result, a normal double bond will form between the two atoms (Hasanuzzaman et al., 2013). Oxygen, which already has lone pairs of electrons, will supply the two missing electrons for carbon. Thus, the third covalent connection will be a coordinate covalent bond.

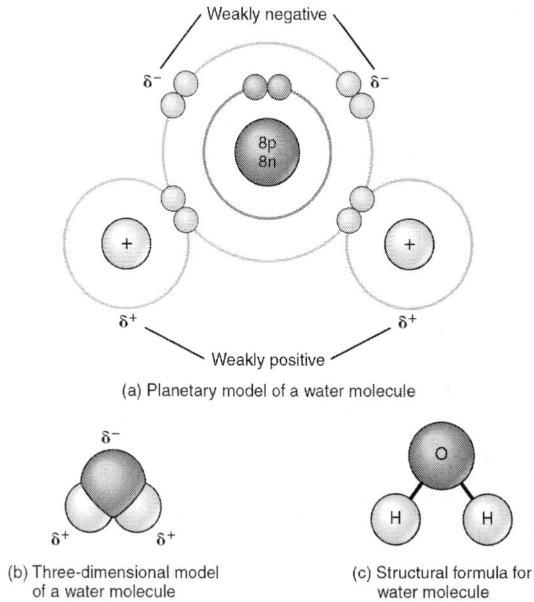

Figure 7.8: The covalent bond in water and oxygen.

Source: https://upload.wikimedia.org/wikipedia/commons/9/99/209_Polar_Covalent_Bonds_in_a_Water_Molecule.jpg

This category includes -bonding, -bonding, metal-to-metal bonding, agostic interactions, bending bonds, three-center two-electron bonds, and three-center four-electron bonds. The term "covalent bond" was used in 1939 to describe the connection between two atoms. The term "covalent bond" in the context of valence bond theory refers to the fact that the atoms involved share the same "valence," which is signified by the prefix co. Covalent bonding allows the hydrogen atoms in the H_2 molecule to share two electrons. When the electronegativity of all atoms is the same, the maximum Covalency is obtained. Instead of both atoms having the same electronegativity, covalent bonding demands that their electronegativity be equal. Electrons in a delocalized covalent bond are spread throughout more than two atoms. There are instances when a molecule's electron configuration and empirically observable features demand a superposition of forms. Because the same two atoms are bound in different Lewis structures, the bond order in these compounds is not an integer (a single bond in one, a double bond in another, or perhaps none at all). The nitrate ion, for example, may be found in three distinct forms that can be compared (Haidekker &

Theodorakis, 2010). Each nitrogen-oxygen bond is double in one structure and single in the other two, giving each N–O interaction a bond order of 2 + 1 + 1. Aromatici organic compounds with planar rings are more stable and symmetric when the number of electrons is 4n + 2 (n is an integer). Benzene is the classic aromatic compound with six electrons bound together (n = 1, 4n + 2 = 6).

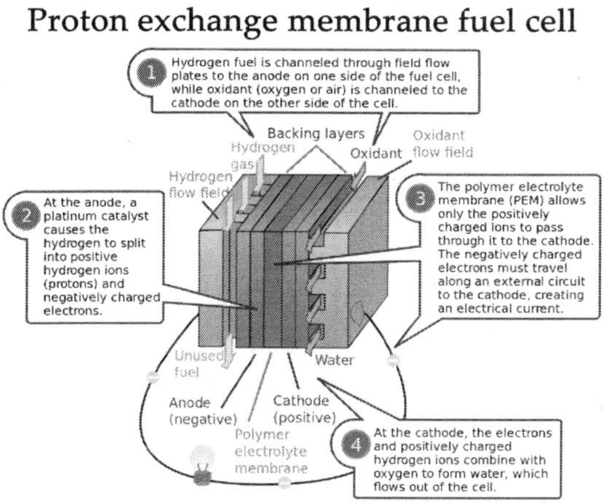

Figure 7.9: The hydrogen fuel cell.

Source: By Jafet at English Wikipedia - Transferred from en.wikipedia to Commons by Kpengboy., Public Domain, https://commons.wikimedia.org/w/index.php?curid=8680999

These molecules occupy three delocalized molecular orbitals or form conjugate bonds in two linearly coupled resonance structures in either molecular orbital theory or valence bond theory, resulting in a more stable regular hexagon than the hypothetical 1,3,5-cyclohexatriene. In heterocyclic aromatics and substituted benzenes, changes in electronegativity across different portions of the ring can impact the chemical behaviour of otherwise equivalent aromatic ring bonds. Hypervalent Some compounds, such as xenon difluoride and sulfur hexafluoride, have greater coordination numbers than would be permitted by the octet rule due to their exclusive covalent bonding. The 3c–4e model examines the molecular wave function using molecular orbital theory and valence bond theory, concentrating on the non-bonding highest occupied molecular orbitals.

When three atoms share two electrons, "3c–2e" three-centre two-electron bonds develop. Boron hydrides, such as diborane (B_2H_6), can show this sort of bonding because they lack the valence electrons required to form localized (2-centre 2-electron) bonds that connect all of the atoms (Vasapollo et al., 2011). Because there are enough bonding orbitals to connect all of the atoms when employing 3c–2e bonds, a newer formulation, molecules may now be categorized as electron-specific. Diborane is composed of two boron-boron bonds that are connected by two electron pairs and one proton-nucleus (the nucleus of the hydrogen atom) transferring electrons. Some cluster compounds have four-center two-electron bonds.

The theory of quantum mechanics has been explained. Following the discovery of quantum mechanics, two key quantum chemical bonding theories were developed: valence bond theory (VB) and molecular orbital theory (MOT) (MO). As a result, a quantum description for atomic contributions to state electronic density is provided. This article compares and contrasts them. Covalent structures come in a variety of sizes and shapes, ranging from molecules to macromolecular structures. In contrast, the attraction between molecules is usually insignificant. Covalent compounds include gases such as HCl, SO_4, CO_2, and CH_4. Inside molecular structures, tiny attractive forces can be detected. Low-boiling liquids (such as ethanol) and low-melting solids (such as magnesium) are examples of covalent substances (such as iodine and solid CO_2). Covalent bonds can be used to connect the atoms of biopolymers such as proteins and proteins, as well as synthetic polymers such as polyethylene. To connect vast numbers of atoms into network covalent structures, sheets (like graphite) or three-dimensional structures (like diamond and quartz) are utilized (or enormous covalent structures).

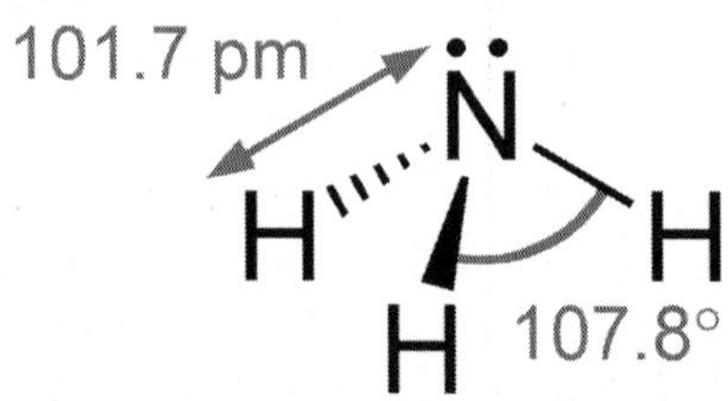

Figure 7.10: Covalent bonding of ammonia.

Source: By Ben Mills - Ain wirk, Public Domain, https://commons.wikimedia.org/w/index.php?curid=6628922

These materials are very brittle and electrically insulating, as well as having exceptionally high melting and boiling temperatures. Massive macromolecular structures are frequently composed of elements with high electronegativity and the capacity to form three or four electron pair bonds. The electron configuration of a molecule can be obtained in one of two ways, as shown by the terms. In valence bond theory, electrons in the hybrid orbitals of atomic compounds are filled with electrons, and then linear combinations of contributing structures (resonance) are performed when there are several contributing structures. In molecular orbital theory, atomic orbitals are linearly merged before electrons are added to the resultant molecular orbitals. This chemical bonding problem can be tackled in a variety of ways. Valence bond theory, which generates the molecular wave function from local bonds, allows for a more precise estimation of bond energy and reaction processes (Antoni et al., 2015). Because it employs delocalized orbitals to construct the molecular wave function, the molecular orbital theory is more suited for calculating ionization energy and interpreting spectrum absorption regions. Both theories provide inaccurate qualitative predictions.

The dissolution of homonuclear diatomic compounds into individual atoms is accurately predicted by Heitler–London valence bond theory, while the dissociation into a mixture of atoms and ions is wrongly predicted by Hartree–Fock molecular orbital theory. Hückel's aromaticity rule can be effectively predicted using fundamental molecular orbital theory and simple valence bond theory, although cyclobutadiene's resonance energy is wrongly anticipated to be larger. Configuration interaction can be used to rectify wavefunctions that are not qualitatively identical or that do not match experimentally observed stabilization energy. To combine the valence bond, molecular orbital ground state functions are merged with functions defining all conceivable excited states utilizing vacant orbitals. To do this, covalent function with all conceivable ionic structure functions. The molecular orbital (MO) approach exaggerates the weight of ionic structures, whereas the basic valence bond method ignores them. This is equivalent to claiming that electron correlation is overlooked by the fundamental molecular orbital approach but overstated by the simple valence bond technique (Turney et al., 2012). Modern quantum chemistry simulations often begin with the molecular orbital technique but progress beyond it, not because the MO method is superior to the valence bond approach, but because the MO method is more suited to numerical computations. Computer computations are more efficient and faster because molecular orbitals are orthogonal compared to nonorthogonal valence bond orbitals.

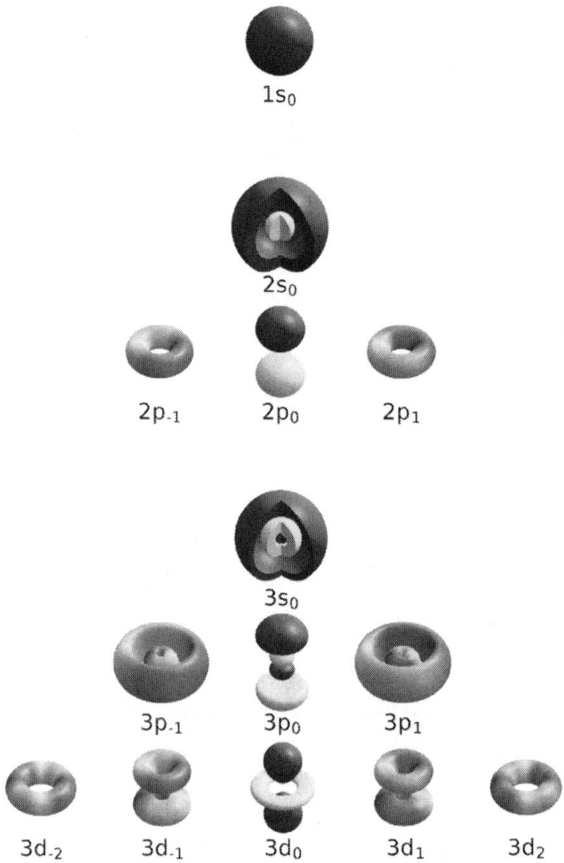

Figure 7.11: Quantum numbers.

Source: By Geek3 - Own work, Created with hydrogen 1.1, CC BY-SA 4.0, https://commons.wikimedia.org/w/index.php?curid=67681892

Covalent bonds far outnumber ionic bonds in organic chemistry. When one or more pairs of electrons are brought to two nuclei at the same moment, a covalent bond is formed. Bonding electrons are those found between two nuclei. Electrons cannot be exchanged between two atoms that form covalent bonds, regardless of how many distinct electronegativity levels they have. Consider the covalent bond that exists between hydrogen atoms. A hydrogen molecule is produced by joining two hydrogen atoms, each with one electron in the 1 s orbital. A covalent bond draws two hydrogen atoms to the same set of electrons. The connection is represented by two dots or a continuous line. Each hydrogen atom absorbs a helium electron state.

When the electrons of two hydrogen atoms create a covalent link, energy is released. Exothermic refers to a procedure that creates heat. At 298 Kelvin, chemical processes creates the heat equal to one molecule of chemical being generated. 435 kJ mole of energy is required to make one mole of hydrogen from two hydrogen atoms (Thomford et al., 2018). Because of the energy wasted during the process, the hydrogen molecule is more stable than the two hydrogen atoms. Removing the two hydrogen atoms that are bonded together needs 435 kJ mole1, also known as the H-H bond strength. When two hydrogen nuclei collide, they create a bond. Attractive and repulsive forces balance out, and separation occurs. Mutual attraction exists between two nuclei and two electrons.

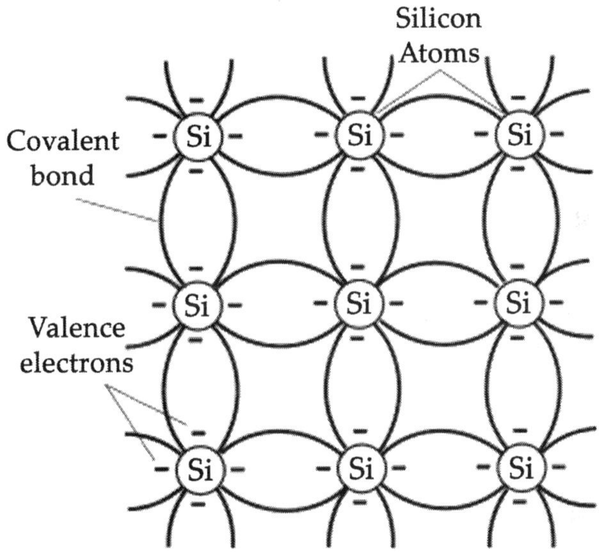

Figure 7.12: Molecular compounds.

Source: https://commons.wikimedia.org/wiki/File:Covalent_bonding_in_silicon.svg

Covalent bonds are the most common type of bonding found in organic chemistry. When two atoms share electrons, a covalent bond is established. When two atomic orbitals cross over one another, a bond is created. To conform with the octet rule, the valence shells of both atoms will get credit for each shared electron. Each atom provides one electron to the electron pair of a single bond. Triple bonds, as opposed to double bonds, have three pairs of electrons, whereas double bonds have two. Each electron pair in a covalent bond is shared by a greater number of additional electrons. Covalent bonds are formed when the electronegativity differences between two atoms are too small to enable electron transfer (Aramwit et al., 2010). Bonding electrons are electrons that exist between two nuclei. The bound pair is the "glue" that holds atoms together in molecular units. The most fundamental covalently bound substance is hydrogen. Each of its two hydrogen atoms may have an electron in the 1s orbital. The two electrons in the covalent bond are shared by both hydrogen atoms, resulting in an electron configuration similar to that of helium. In the majority of stable neutral compounds, hydrogen and halogen atoms form a single covalent bond with other elements. Carbon, oxygen, and nitrogen atoms, on the other hand, can bind to a high number of atoms at the same time.

The valence of an atom is related to the number of these bonds. Nitrogen has three valences, carbon has four, and oxygen has two. A covalent bond connects two atoms and an electron pair. Other ligands, such as sulfate, amide, silane, carboxyl, and hydroxyl groups, are covalently connected to the MONPs. These ligands serve as connecting agents between biomolecules and nanoparticles. Covalent carboxyl groups can be employed to attach mesoporous SiO_2 NPs to PEG or PEI, for example. Cisplatin was also integrated into SiO_2 NPs coated with PEG or PEI, and the drug's subsequent release was investigated. PEI-coated SiO2 nanoparticles outperformed PEG-coated SiO_2 nanoparticles in terms of cisplatin administration (Shin et al., 2019). We immersed surface-functionalized ZnO NPs in solution for 24 hours using direct mixing and continuous stirring. FTIR and Raman spectroscopy were used to assess surface functionalization. Depending on the findings, a single or two Si-O-Zn (covalent bond) bonds were established between APTES and the ZnO NP surface.

CHAPTER 8

MOLECULAR STRUCTURES

CONTENTS

8.1 INTRODUCTION

Structural equations and molecular models may be used to explain the electronic structures of molecules; however, this is only possible if it is also practical to calculate the electron orbitals that are occupied by the atoms that make up the molecule. The method of determining a molecule's structure may be utilized on a diverse collection of substances, ranging from those that are the most fundamental (diatomic oxygen or nitrogen) to those that are the most complex (organometallic complexes) (e.g., protein or DNA).

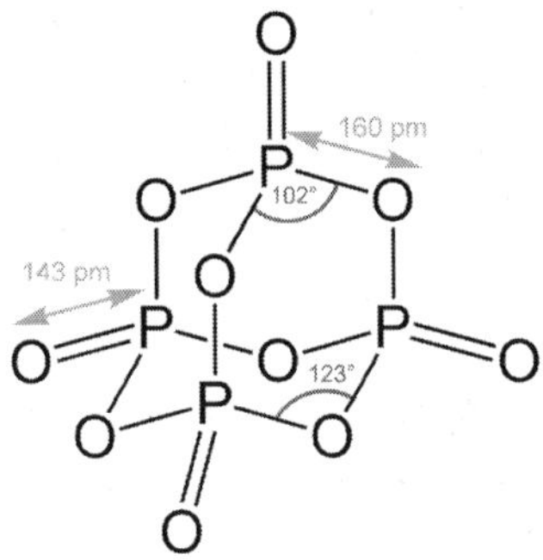

Figure 8.1: Molecular Structure of Phosphorus pentoxide.

Source: Public Domain, https://commons.wikimedia.org/w/index.php?curid=1397245

Around the year 1858, several scientists, including August Kekulé, Archibald Scott Couper, and Aleksander Butlerov, developed hypotheses on the composition of chemical compounds (Aidas et al., 2014). These ideas, which were the first to establish that the valency of atoms inside molecules governed the precise sequence of chemical compounds, may be used to either find or solve a structure in three dimensions. This was a groundbreaking discovery in the field of molecular chemistry, and it completely shifted the paradigm.

In terms of the chemical composition, a three-dimensional arrangement that is referred to as the molecular configuration, and an accurate representation of (relative) atomic coordinates, which includes information such as chirality, one can make a significant distinction between the three. This distinction is important. Establishing the structure of chemical compounds serves several purposes, the most fundamental of which is to determine the kind and degree of bonding that exists between all of the atoms that are contained inside the molecule. During this process, it is also envisaged that the spatial coordinates of the atoms in all three dimensions would be able to be determined (or other solid). The following is a list of methods that may be used to figure out the structure of a molecule:

- The atomic-to-atomic connection is the only thing that matters for proton and carbon-13 measurements in NMR and other mass spectrometry techniques (to give overall molecular mass, as well as fragment masses).
- Inferential methods, such as absorption spectroscopy and vibrational spectroscopies, such as infrared and Raman spectroscopies, are both examples of alternative approaches to get insight into the electronic structure of molecules. These spectroscopies are both examples of spectroscopies (Shi et al., 2010).
- Performing all measurements using the extremely accurate metric system Information on the three-dimensional structure of gases can be obtained by techniques such as gas electron diffraction, microwave (rotational) spectroscopy, and other rotationally resolved spectroscopies. X-ray crystallography and neutron diffraction are two techniques that have the potential to deliver information on crystalline materials in a three-dimensional format.

These approaches have the potential to build three-dimensional models with atomic precision, which translates to length increments of 0.001 millimetres and angle increments of 0.1 degrees (in unusual cases even better).

Electron-nuclear double resonance (ENDOR) and electron-spin resonance spectroscopes (ESRS) are two methods that may be utilized to get additional information in situations in which the functional group of a molecule possesses an unpaired electron spin. When a molecule contains metal atoms, the use of NMR and crystallography becomes increasingly significant since neither crystals nor the specific atom types required by NMR are conveniently accessible for structure identification. For instance, electron microscopy is a technique that may be utilized under specific conditions.

8.2 DETERMINING MOLECULAR STRUCTURE

Figure 8.2: Determining molecular structure: Bruker 700 MHz nuclear magnetic resonance (NMR) spectrometer.

Source: By Mike25 - Own work, Public Domain, https://commons.wikimedia.org/w/index.php?curid=4127397

The process of finding the nuclear centers of a molecule can make use of a variety of techniques, including neutron diffraction and microwave vibration-rotation spectra, to name just two of the possibilities. Neutron diffraction is the name given to each of these methods of analysis. A procedure known as X-ray diffraction is a technique that can be applied for investigating the electron cloud that is situated around the nucleus of a molecule (Arif et al., 2010). This investigation may be carried out. The group of more knowledge can be accomplished through either the technical procedures of electron spin resonance or nuclear magnetic resonance. The advent of electron microscopy made it feasible, for the very first time, to investigate individual molecules and atoms in the context of their native surroundings.

To hypothetically compute the molecule structure, the Schrodinger equation, which is a quantum mechanical equation that describes the movement of electrons in the presence of nuclei and is used in the process, is applied (a quantum mechanical equation for electron transport in the field of nuclei). These are the bond lengths and angles that, in terms of the structure of the molecule, create the least amount of molecular energy that is practical. This is because these bond lengths and angles produce the least amount of molecular energy. Scientists are currently employing computers and supercomputers in an attempt to numerically solve the Schrodinger equation in the pursuit of discovering new structures. This enables them to recognize novel patterns more easily.

The positive and negative charges that are connected with a molecule will cancel each other out if there is no overall net electrical charge associated with the molecule. The method in which positive and negative charges are spread over the expanse of space is what determines the kinds of forces that such molecules are subjected to. If the internal structure of a molecule can be defined as having spherical symmetry, then that molecule is said to have a nonpolar structure.

A measurable propensity to rotate in an electric or magnetic field is referred to as the dipole moment. This propensity is also known as the inclination to spin in a field that is either electric or magnetic. Because of this, a dipole moment is said to have polar characteristics if it possesses an excess of positive charge on one end and an excess of negative charge on the other end. When polar molecules can spin, they tend to orient themselves in ways that produce attractive forces for the environment that they are surrounded by (Barbatti et al., 2014). Because the positively charged and negatively charged amino acid residues that make up its composition are arranged in a complementary manner, ionic-complimentary peptides have a distinct chemical structure. This is because ionic-complimentary peptides have a distinct chemical structure.

The ordered charge sequence is what causes one-of-a-kind electrostatic interactions, which, in turn, regulate the self-assembly of molecules. These interactions are what make the self-assembly of molecules possible. The more well-known hydrogen bonds and the van der Waals forces are both examples of these types of interactions. It is necessary for the distribution of the charges on the molecules to take on a specific shape to make it easier for the molecules to assemble themselves through the process of ionic complementarity. This is because it is easier for the molecules to assemble themselves when the charges are organized in a particular way.

The categories of ionic-complementary peptide sequences that are denoted with the descriptions I, +, II, ++, and IV, ++++ have garnered the most interest from researchers. As a result of this, the positions occupied by positively charged amino acid units and negatively charged amino acid units interchange places within groups of one, two, or four. This is an immediate consequence of the predicament. If you connect these combinations or repeat them to build additional ionic-complementary peptides, you can end up with more sophisticated self-assembly. This is anything that can be done to accomplish the objective.

When compared to surfactants, which have a "head-to-tail" shape, self-assembling ionic-complementary peptides have an amphiphilic structure with alternating hydrophobic and hydrophilic amino acid residues. This is in contrast to surfactants, which have a structure that is not amphiphilic. Because of the way this structure is constructed, 'side-to-side' interactions are possible. For instance, when looking at a -strand peptide, there are hydrophobic residues on one side of the peptide backbone, and hydrophilic residues on the opposite side of the peptide backbone. Both ionic-complimentary peptide self-assembly and the self-assembly of surfactant micelles occur simultaneously; nonetheless, they are two separate processes that may be seen (Sneskov & Christiansen, 2012).

EAK16-II is a 16-amino acid peptide segment that was synthesized from the Z-DNA-binding protein. It has been the impetus for studies into a family of ionic-complementary peptides that were not previously recognized. EAK16-I and EAK16-IV, two variations of this peptide, are members of the type II group; nevertheless, you may find them in this peptide as well. The peptides that are a part of this family can self-assemble into fibril structures that contain a significant amount of sheet material. The hydrophobic side of the peptide is responsible for hydrophobic interaction, electrostatic contact, and hydrogen bonding, all of which contribute to the realization of this possibility through their cooperative efforts. Researchers have developed new ionic-complementary self-assembling proteins as a result of an in-depth investigation into the molecular structure of EAK16-II. For example, the peptide RADA16-I was developed to improve the biocompatibility of tissue scaffolds as well as cell adhesion.

Ionic-complementary peptides can have a broad variety of different molecular shapes and functionalities because the charge distribution and chain length of these peptides can vary, and amino acids with similar properties can be exchanged for one another. Both the distribution of charge and the order in which it is provided can affect the secondary structure and the ability of a system to assemble itself. The length of a peptide is a key factor to consider when assessing whether or not it is compatible with partnering with other molecules. Peptides of type I require four amino acids, peptides of type II require eight, and peptides of type IV require sixteen amino acids. Third, the ionic-complementary peptide family may be expanded by exchanging certain amino acids with one another (Stuppia et al., 2012). There are several routes that one might use to achieve this goal. The hydrophobic residue alanine (A) in EAK is changed with either phenylalanine (F) or leucine, which results in the formation of EFK and ELK (L). Even the most minute changes do not have an impact on two of the most fundamental properties, namely, ionic complementarity and self-assembly.

8.3 VSEPR THEORY

Deducing the molecular structure from the Lewis structure, which is a representation of the molecule's electronic configuration and includes the number of bonds and lone electron pairs, is possible. The molecular structure of a molecule includes the predicted bond angles around the central atom. The Lewis structure of a molecule is a representation of the molecule's electronic configuration. The mutual repulsions between electron pairs in the valence shell of a central atom should diminish, by the Valence shell electron pair repulsion or VSEPR model, as the electron pairs spontaneously organize themselves in a more stable configuration. At the same time, there is an increase in the distance that separates the electron pairs. This is the underlying way of thinking that forms the basis of the notion (Rhoads et al., 2012). This would be the scenario that would play out in the case that they were successful in overcoming the mutual enmity that they felt toward one another. Two distinct types of electron pairings can take place in the valence shell of an atom.

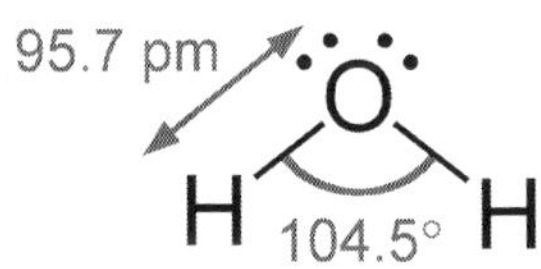

Figure 8.3: VSEPR Theory: Example of bent electron arrangement. Shows the location of unpaired electrons, bonded atoms, and bond angles. (Water molecule) The bond angle for water is 104.5.

Source: By Ben Mills - Own work, Public Domain, https://commons.wikimedia.org/w/index.php?curid=6834720

Both of these types of electron pairings are possible. There are also solitary pairs and bonded pairs present here. Atoms that are already coupled to one another have an increased likelihood of running into atomic pairs that are travelling alone. It is possible to minimize the amount of electrostatic repulsion that occurs between electrons by breaking up several places that have a very high electron density. This may be done by dividing up many regions that have extremely high electron densities.

The VSEPR theory is capable of accurately predicting, in an extremely high percentage of situations, not only the arrangement of electron pairs surrounding each central atom but also the organization of atoms inside a molecule. This is because the VSEPR hypothesis was proposed in the 1960s, which is the reason why this is the case. If one were to give it some thought, the VSEPR theory may allow for the possibility of hybridization. Nevertheless, this theory can only explain the interactions that take place between pairs of electrons at a time. This is not something that can be ignored under any circumstances (Rastogi & Sani, 2011). The ultimate arrangement of atoms inside the structure of a molecule is determined in large part by other interactions, such as nuclear-nuclear repulsions and nuclear-electron attractions, both of which play an essential role in the process. Nuclear repulsions and nuclear attractions are the two categories that may be used to classify these interactions with one another. Depending on how you look at it, these interactions can be put into either the category of nuclear repulsions or nuclear attractions.

The VSEPR theory is built on several fundamental presumptions, some of which are listed below for your perusal: The electrons in the valence shell are less attracted to one another than the other pairs of electrons. This is because the electron clouds in the valence shell are negatively charged. In comparison to the other pairs of electrons, this one is unique. These electrons make an effort to occupy space in a manner that will maintain the distance between them as short as possible. This is done so that they would not be repulsed by one another when they come into contact with one another.

Studying the elements that make up a molecule of gaseous beryllium fluoride can provide a particularly illustrative example of how the VSEPR theory operates. The Lewis structure of the molecule of beryllium fluoride reveals that the beryllium atom at the nucleus of the molecule is only surrounded by two electron pairs. This can be seen by looking at the molecule. Because there won't be any lone electron pairs present when there are two bonds present, the core atom won't have any lone electron pairs because there won't be any lone electron pairs present (Raies& Bajic, 2016). Another benefit of this arrangement is that it reduces the amount of electrostatic repulsion that occurs between electrons when they are on opposite sides of the core atom rather than when they are together. Once the acute angle reaches 180 degrees, a connection has been made between the two points.

The molecular structure of an atom and the geometry of its electron pairs are two distinct concepts that are relevant here. Both concepts are at play. The molecular structure of a substance is not determined by the distribution of electrons within that substance; rather, the molecular structure is established by the arrangement of the atoms that comprise that substance.

To differentiate between these two possible outcomes, the geometry of the electron pairs, which includes all of the electron pairs, will be utilized. The only structure besides the molecular structure that contains information on the positions of atoms within a molecule is the molecular structure; all other structures are completely devoid of this information. The geometries of the electron pairs and the structures of the molecules will be comparable if there are no lone electron pairs near the core atom. When there are lone pairs on the core atom, on the other hand, the geometries of the electron pairs will appear to be different from one another.

Methane, which has the chemical formula CH_4 and is distinguished by having four pairs of electrons attached to the carbon atom at its core, makes up the majority of natural gas. Methane can be identified by its chemical formula. The chemical formula for methane is CH4, and some of the characteristics that set it apart are as follows: The tetrahedral geometry seen in the electron pairs is also replicated in the molecular structure of methane. The electron pairs in NH3 take on a tetrahedral shape (Pritchard et al., 2019). This is because there are four electron pairs linked to the nitrogen atom. This is where the name of the molecule comes from. To refer to something by such a name requires a certain amount of intestinal fortitude. One of these sections in particular does not contribute to the chemical structure of the molecule, but it does change the shape of the molecule; this is true for all of these sections, not just one of them. Even though lone pairs do not contribute to the structure of the molecule, they do change the shape of the molecule.

8.4 MOLECULAR STRUCTURE: THE FIVE BASIC SHAPES

We can utilize Lewis structures to make predictions about a molecule's three-dimensional structure once we've successfully drawn them for simple compounds. After drawing Lewis structures for a few simple compounds, we're ready to go on to the next step. Predictions of this nature can be aided by the information provided by Lewis structures. Because of the VSEPR model, or valence shell electron pair repulsion, it is possible to achieve this result.

The steric number (abbreviated as SN) of a molecule's central atom is not an easy number to decipher. SN can also be used to refer to this number. The term "electron density" is sometimes used to describe the steric number. To find the total number of bound atoms, add the lone pair number of atoms to the total number of atoms. The total number of atoms that are attached to the core atom can be found here. Although we use the term "steric number," Tro refers to it as the "number of electron groups" in his book (Tro NJ. Chemistry: Structure and Properties, 2nd ed). Tro's "electron groups," which correspond to electron domains, are used far more frequently than the phrase "electron pair" in calculating SN (Bacellar et al., 2015). Since we only count a bound atom once, even though it may be double or triple bonded to the core atom, this is the case because we only count bound atoms once, this is the reason for the discrepancy. The total number of electron domains has a directly proportional link to the value of the SN, which is defined by this relationship (electron groups).

If an atom follows this rule, the total number of neutrons it contains cannot be more than four at any given time because of a law called the octet rule. Because of the regularity with which this takes place, we will start our inquiry by concentrating on the many potential chemical configurations for the steric numbers 2, 3, and 4.

8.4.1 Trigonal Planar Geometry (with SN equal to three)

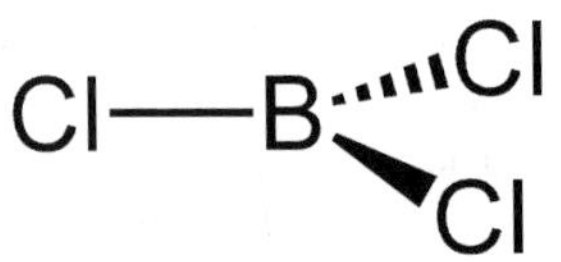

Figure 8.4: Trigonal Planar Geometry (BCl3).

Source: By Ben Mills - Own work, Public Domain, https://commons.wikimedia.org/w/index.php?curid=973385

Several molecules, including boron trichloride (BCl_3), formaldehyde (CH_2O), and carbonate anion, are examples of the geometry known as trigonal planar geometry. This geometry is identified by an SN value of

3. (CO_3^{2-}). The VSEPR model predicts that the electron groups will be placed in a trigonal planar geometry, and the four corners of an equilateral triangle will serve as the alignment points for the electron groups. This is a fundamental assumption that the model is based on. The factual basis for this concept is found in the observation that electrons typically cluster in groups of three (Pinzi& Rastelli, 2019). When every electron group works as a bonding group, a bond will have an angle of exactly 120 degrees, which is believed to be the optimal angle for a bond to have. In other words, when every electron group functions as a bonding group.

8.4.2 Tetrahedral-based Structures (SN equal to 4)

Tetrahedral-based structures are those that have four sides and are shaped like a cube. In the scenario in which SN is equal to four, the structures of the tetrahedral form have been seen. As a result of the four-electron domains coming together and cooperating, the atoms that are now linked to one another take up positions in the four corners of a tetrahedron (then there are four atoms attached to the centre atom). The angles, which add up to a total of 109.5 degrees, allow for the best possible bonding to be achieved as a consequence of this design.

The angle created by any two C-H bonds is 109.5 degrees, which is the same as the bond angles for methane. In other words, the angle generated by any two C-H bonds is the same as the angle formed by methane. The perfect tetrahedral angle is preserved in a molecule whenever there are four atoms of the same sort attached to the carbon atom in the middle of the molecule. This particular angle is referred to as the tetrahedral angle. When one of the bonding electron pairs is switched out for a lone pair of electrons that do not create bonds, the bond angles for the remaining bonds tend to deviate from their ideal values (Patil et al., 2008). This happens because the lone pair of electrons do not participate in the formation of bonds. This takes place as a result of the fact that the lone pair of electrons do not contribute in any way to the creation of bonds.

The negative charge of the lone pair is more widely dispersed in space, which repels the bonding pairs (which are more constricted, being confined to the area between the two bound atoms), causing them to move closer to one another in space. Because the bonding pairs are more constricted, they are confined to the area between the two bound atoms. Because the spaces between the bonding pairs are smaller, the bonding pairs are constrained to the region between the two atoms that are linked together. Bonding pairs are restricted to the region between the two atoms that are connected because the intervals between the bonding pairs are smaller. This effect is seen when moving from methane (four bonding pairs, zero lone pairs, bond angles 109.5°) to ammonia (three bonding pairs, one lone pair, bond angles 107.3°) to water (two bonding pairs, two lone pairs, bond angles 104.5°). Methane has four bonding pairs, while ammonia has three bonding pairs and one lone pair. Methane contains four different pairs of bonding electrons, but ammonia only has three different pairs of bonding electrons and one lone pair (Ogino et al., 2013). Ammonia, on the other hand, only possesses three bonding pairs, whereas water only possesses two bonding couples in addition to two lone pairs. Methane is the element with the greatest number of different bonding partners.

We can refer to the geometry of the electron domain as "that of the molecule." The underlying geometry of each electron domain is tetrahedral as a result of the fact that each scenario has a total of four electron domains. This is a direct result of the fact that each electron domain possesses a total of four electrons (Reminder: The number of electron domains for a central atom is the same as its SN or a steric number and is always the number of bonded atoms plus lone pairs). On the other hand, one can discern the structure of the molecule when only the atoms are taken into consideration and there are no lone pairs present in the system.

The underlying electron domain geometry of methane is tetrahedral. This is the case because each electron domain that surrounds the core carbon atom corresponds to a hydrogen atom that is covalently bound. This holds for both the structure of the molecule itself as well as the geometry of the electron domain. This is because methane is a chain that consists of four different elements altogether. The chemical structure resembles a trigonal pyramid when one of the tetrahedron's vertices is removed, and when the shape created by the NH_3 molecule's four atoms is inspected. This is because there are four atoms in the NH_3 molecule (Bennett et al. 2010). This occurs as a result of the fact that each molecule of NH_3 is comprised of four atoms (a pyramid with a triangle base). The water molecule is composed of only three atoms, and it can take on two distinct molecular shapes: linear, in which the three atoms lie in a straight line with a bond angle of 180 degrees, and "angular," in which the atoms do not lie in a straight line and the molecular shape is "bent," also with a bond angle of 180 degrees. Both of these molecular shapes have the same bond angle of 180 degrees.

These two distinct possibilities for the structure of molecules share the same bond angle of 180 degrees. In addition, a ring composed of three members may be found inside the water molecule structure. Given that the underlying electron domain geometry of the water molecule is tetrahedral, our best estimation for the value of the bond angle is that it will be somewhere in the range of 109.5 degrees. This is based on the fact that the water molecule has a tetrahedral electron domain structure. Previously, it has been established that the lower angle observed is due to lone pair attraction on the oxygen atom in the core balancing out. There is now a clearer understanding of why the reported angle was so much lower than predicted. This explains why the actual measurement and reporting showed a much shallower angle.

8.4.3 Triangular Bipyramid Bases Structure (SN equal to 5)

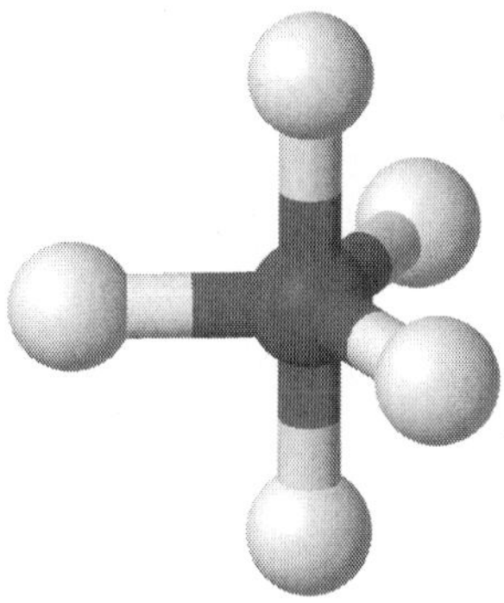

Figure 8.5: Triangular bipyramid bases structure.

Source: Public Domain, https://commons.wikimedia.org/w/index.php?curid=1454616

Triangular bipyramid bass structure are found in five different configurations that are comparable to the structure of the standard nucleotide (SN). This type of structure is classified as a "trigonal bipyramid" when the SN value is more than or equal to 5. There are four possible places for the atoms that are joined to form a trigonal bipyramid in which all five electron domains are linked to one another: (then there are five atoms attached to the central atom).

8.4.4 Octahedral Geometry (SN = 6)

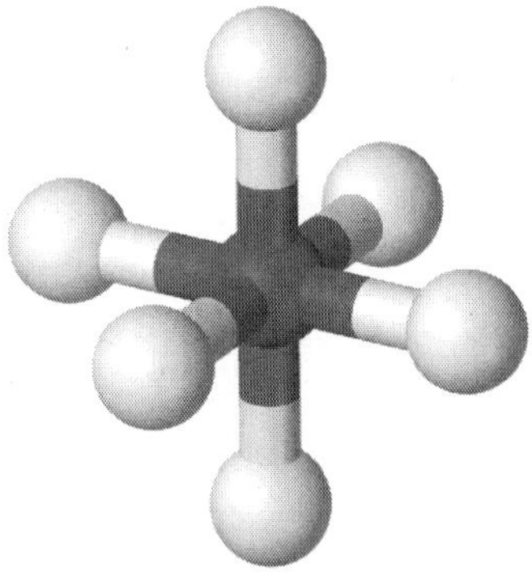

Figure 8.6: Octahedral geometry.

Source: Public Domain, https://commons.wikimedia.org/w/index.php?curid=1454714

Six-symmetry designs are those made utilizing octahedral geometry and having six symmetries. The term "octahedral geometry designs" refers to these sorts of patterns. One way to express this is using the formula SN = 6 for the structure. Some sulfur hexafluoride, bromine Penta fluorides, and xenon tetrafluorides (all of which include this structure) may be found in nature Molecular geometries are the structures that arise as a result of examining how bound atoms are structured in space (despite discounting the steric effects of lone pair interactions), and these compounds reveal a broad diversity of molecular geometries (O'Gorman et al., 2012). Molecular geometries are the forms that arise as a result of understanding how bonded atoms are structured in space. The forms that become visible as a result of understanding how atoms that are linked together are organized in space are referred to as molecular geometries.

Once we have a comprehensive knowledge of the chemical structure that a molecule is likely to comprise, we will be able to establish whether or not a molecule possesses a net dipole moment. A net dipole moment may be determined using this technique. A bond dipole will be produced in the succeeding connection if the electronegativity of the atoms that are linked together suffers a change after they have been connected. Before continuing with the study, it is required to identify whether or not a structure has a strong bond dipole. If it does, then go to the following step. When attempting to determine the molecule's total net dipole moment, it is crucial to think of the bond dipoles as vectors and to have a grasp of how to put them together (Neese, 2012). This may be performed by taking into mind the following: Since they interact with identical molecules, bond dipoles are often removed. As a result of their interplay, this is the case. Carbon dioxide's highly polar: C=O bonds serve as an excellent illustration of this principle. One of the best examples of this idea is carbon dioxide. Two bond dipole moment vectors are generated as a result of these bonds, each of which has the same length but faces in the opposite direction. These interactions cancel each other out, resulting in a bond angle equal to 180 degrees due to the linear geometry of CO_2 molecules.

CHAPTER 9

GASES AND THE ATMOSPHERE

CONTENTS

9.1 INTRODUCTION

Air is a mixture of gases and is constantly moving. It's an essential part of our world and without it, life would not be possible. Air supports our breathing, growth, and many other biological processes. It also makes up most of Earth's atmosphere which consists mainly of nitrogen (78%) and oxygen (21%). The remaining 1% consists of argon, carbon dioxide, neon, helium and other trace gases that are important for life on earth to exist (Nwodo et al., 2012).

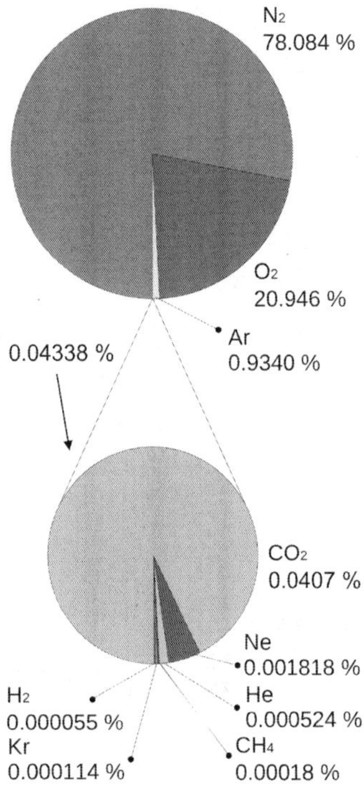

Figure 9.1: Composition of gases in the atmosphere.

Source: By Mysid - Vectorized version of w:Image:Atmosphere gas proportions. gif (originally by Brockert). I SVG'd it a) to make it more international (chemical symbols) b) to make it (hopefully) clearer when rendered as a thumbnail and c) to make it easier to modify., Public Domain, https://commons.wikimedia.org/w/index.php?curid=823817

The flow of the upper atmosphere carries gases around the world. This flow takes place in four different cells: the tropical and subtropical jet streams, the polar jet stream, and the mid-latitude westerlies. The atmosphere consists of primarily nitrogen (78 percent), oxygen (21 percent), argon (0.93 percent), carbon dioxide (0.03 percent) and other trace gases. The atmosphere is always in motion, transporting water vapor and gases around the world. The mixture of gases and particles in the atmosphere is constantly moving (Moseti et al., 2016). As the Earth rotates, air masses from different regions of the globe continue to mix, changing temperature and pressure. Wind is one way in which atmospheric gases and particles move around. The movement of air creates currents, which in turn create weather. The composition of air, which is constantly changing as the gases that make up our atmosphere are continually being exchanged and recycled.

The word "atmosphere" comes from the Greek word for aroma, because ancient Greeks believed that air contained some sort of life force. Today we know that air does not contain any life-sustaining gases and that our bodies, like all forms of matter, are made up mostly of inert materials like carbon and hydrogen. Yet the atmosphere is essential to life on Earth. By allowing solar radiation to pass through while blocking harmful ultraviolet radiation, which can damage living things, it is so vital to all living things.

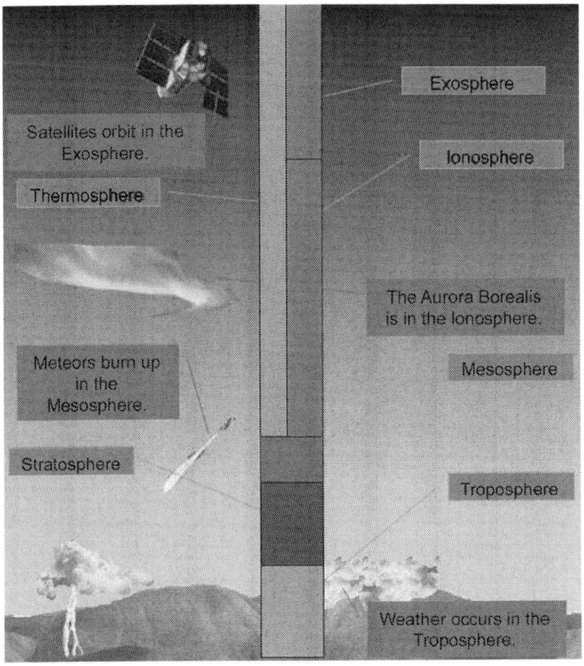

Figure 9.2: The various layers in the atmosphere.

Source: https://en.wikipedia.org/wiki/File:Layers_of_the_atmosphere.jpg

Our atmosphere consists of all of the gases surrounding the Earth, excluding the air we breathe. Earth's atmosphere comes in two layers called the troposphere, which accounts for 75% of our atmosphere and extends up to about 10 miles above the Earth's surface, and the stratosphere, which can reach far higher into space but accounts for a considerably smaller portion of our planet's overall composition.

In a nutshell, the atmosphere is the gaseous envelope that surrounds a planet. It does not extend beyond the orbit of the Moon, and for most planets (with the notable exception of Jupiter), it contains approximately 80% hydrogen and 20% helium by mass, with smaller amounts of other elements including oxygen and nitrogen (Congrains et al., 2013). The atmosphere protects life on Earth by absorbing ultraviolet solar radiation. This process also creates clouds, which can make water available to plants by making rain.

The atmosphere is defined as the layer of gases surrounding the earth. The atmosphere protects life from harmful ultraviolet rays from the sun. It also keeps the temperature of earth relatively stable. The atmosphere is an important natural resource that supports life on Earth. It protects the Earth from harmful radiation, moves water and nutrients around the planet, holds all living organisms in place, and moderates global temperatures. The gases that make up the atmosphere have an average pressure of 14 pounds per square inch (psi) at sea level. It also protects us from meteorites and other dangerous objects that might collide with our planet. Deposits in the form of ice at high altitudes reflect sunlight, reducing heating. Water vapor helps trap heat near Earth to keep temperatures constant. Without a healthy atmosphere, we would not be able to breathe.

Some of the problems faced by the earth atmosphere include greenhouse gases which includes majorly nitrogen, oxygen and a small percentage of carbon dioxide, contamination from other harmful gases, depletion of the ozone layer due to the harmful ultra-violet rays and contamination by various particles and dust in the earth.

As a matter of fact, the atmosphere is a complex system that continually evolves in reaction to both natural processes and human activity. This chapter describes the main features of the atmosphere, climate, and air pollution (Campos & Vasconcelos, 2010). It analyzes how gases interact with each other, with particles in aerosols, and with living things; how sunlight affects the atmosphere; how winds transport heat and moisture; and more.

9.2 FEATURES OF THE ATMOSPHERE, CLIMATE AND AIR POLLUTION

The atmosphere contains all of the gases found on Earth that can absorb and scatter radiation. Of these, water vapor, oxygen and nitrogen are by far the most important in terms of their abundance and radiative effects, but other components like carbon dioxide also play important roles. Carbon dioxide is particularly important because it absorbs infrared radiation emitted from the Earth's surface, warming the lower atmosphere (surface up to about 500 m) through what is known as greenhouse gas effect. The gases in the atmosphere are held in place by Earth's gravity. The weight of all this gas exerts a powerful pressure that keeps us from bursting out into space. This weight also compresses our atmosphere, making it denser near the surface than at higher altitudes.

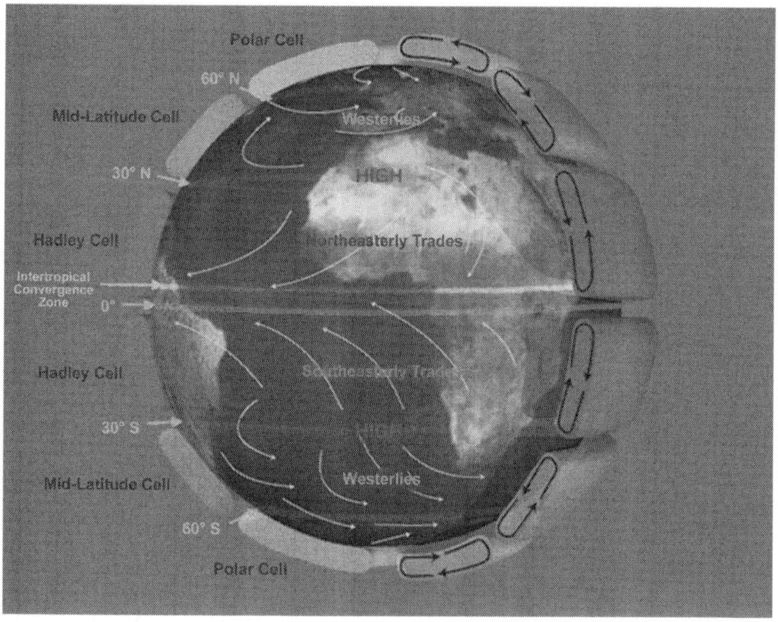

Figure 9.3: The basic drivers of climate surrounding the atmosphere.

Source: By NASA - File:Earth Global Circulation.jpg, Public Domain, https://commons.wikimedia.org/w/index.php?curid=81412461

A variety of climatic conditions may be found in the atmosphere. These include cyclones, anticyclones, hurricanes, and monsoons. The distribution of these climatic variables depends on the location on Earth.

In the atmosphere, the temperature is constant across layers at all locations in a vertical column of air. The temperature profile of an atmospheric layer shows the vertical distribution of temperature averaged over time and space, it could be using average hourly observations taken over a whole month or twenty-four hour averages taken on different days at various places (Colussi et al., 2013). High pressure or anticyclone conditions occur when smooth, unbroken lines of atmospheric pressure form around the center of a large empty area; low pressure or cyclone conditions occur when high winds and storm clouds stir up rough areas that disrupt the circular pattern of weather patterns. With rising air molecules moving away from Earth's surface (law of convection), air moves faster at higher altitudes and is thinner than lower down where friction with Earth's surface slows it down. Being warmer than other layers above it (law of thermal expansion) makes it expand; this lowers its density and increases its buoyancy. Warm and cold fronts, high pressure areas and low-pressure areas are all decided based on the location of warm and cold air masses in the atmosphere. Air masses are large blocks of air that move with the same temperature range, and which can be observed at a variety of distances from Earth's surface. Cold air masses originate near Polar Regions, while warm air masses form during the summer months in middle latitudes.

The atmosphere, along with the hydrosphere, biosphere and geosphere make up the major systems of our planet. The temperature of the atmosphere regulates the overall climate of Earth. Temperature affects weather, which in turn affects both human and ecological events. As a matter of fact, the atmosphere consists of several layers, each with a different combination of physical properties. These layers are distinguished by their height and varying temperatures, pressure, density and chemical composition. The troposphere is the lowermost layer of the atmosphere and is where we live (Chen et al., 2015). Above it lies the stratosphere, mesosphere, thermosphere and exosphere. Air masses are horizontal extensions of air at one level or another that are large enough to move around weather systems. The water cycle involves the continuous exchange of energy, matter and moisture between Earth's surface and the atmosphere by evaporation, condensation and subsequent precipitation.

Air pollution within the atmosphere includes the emissions of reactive gases and particles into the air. Once in the atmosphere, these particles spread out and create tiny airborne dots (particulate matter), where upon they are picked up by wind gusts and pushed inland to other countries or continents. The emission of carbon dioxide is a major factor leading to pollution of air. Carbon dioxide is a gas produced naturally by animals and is an essential part of life on Earth, as plants need CO2 to grow. It is also a product of coal-burning power plants, cement factories and steel mills. Carbon dioxide is accounted for about 40% of the greenhouse gases that affect the earth's temperature. Like the other major greenhouse gases, carbon dioxide traps heat in Earth's atmosphere - contributing to global warming (Chaplin, 2006). The ocean, plants and soil uptake some of the carbon dioxide. These combined processes have stabilized atmospheric concentrations at about 450 ppm so far. Other pollutants that contribute to air pollution in the atmosphere include sulfur dioxide, nitrogen oxides and volatile organic compounds (VOCs). Rising world population and increasing standards of living have contributed significantly to the problem.

The burning of fossil fuels such as oil and coal produces carbon dioxide by chemical reaction during combustion with oxygen that releases heat. These fuels are widely used in producing electricity to run factories, operate vehicles, heat homes and power industries. In order to have clean air, we need to control the sources of pollution. Our industry produces toxic gases, which are released into the atmosphere. These almost always combine with moisture and as a result can form acids. Industrial waste is also used for producing harmful chemicals, like chlorine and fluorine which can be pollutants too.

9.3 SOURCES OF GASES IN THE ATMOSPHERE

Did you know that there are a lot of sources of gases in our atmosphere? Some are natural, and some are man-made. Here are some examples:

- Volcanoes

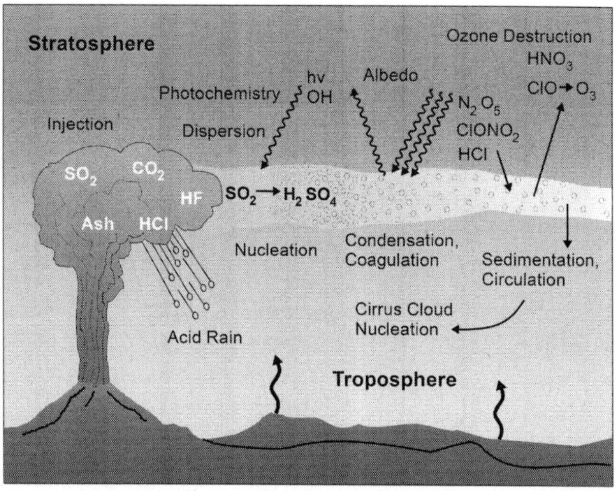

Figure 9.4: An illustration of how volcanoes affect climate and the atmosphere.

Source: Image by USGS (https://www.usgs.gov/natural-hazards/volcano-hazards/volcanoes-can-affect-climate)

Volcanoes are one of the most important sources of gases into the atmosphere. These gases include water vapor and carbon dioxide, which are involved in regulating Earth's climate. Volcanic eruptions release gases such as carbon dioxide (CO2), sulfur dioxide (SO2), and hydrogen sulfide (H2S) into the atmosphere (Milo et al., 2010). These gases dissolve in water droplets to form acid rain, which can have a devastating effect on ecosystems when it falls from the sky. In the atmosphere, volcanic gases are responsible for about a third of all emissions. Volcanoes pose a potential threat to humans, as the gases and volcanic ash released into the atmosphere can damage lungs, impede breathing and cause respiratory disease.

- Biogenic emissions

Biogenic emissions are caused by plants. Trees and other plants release greenhouse gases such as carbon dioxide and methane naturally through a process called respiration. Respiration is the process of using oxygen to convert food into energy, which releases CO_2 and water vapor into the atmosphere. Biogenic emissions are also a part of the carbon cycle, which describes how carbon moves throughout our environment. Plants absorb CO_2 from the air to build their tissues; when they die or decay, they release CO_2 back into the atmosphere as they decompose over time, which contributes to global warming (more on this later).

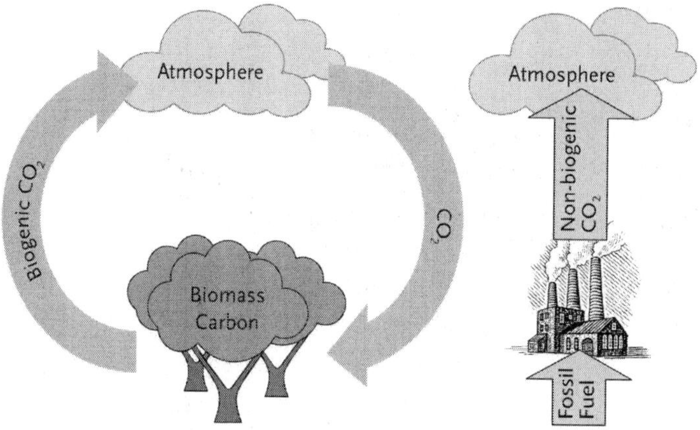

Figure 9.5: Biogenic emissions.

Source: Image by IEA Bioenergy

Biogenic emissions represent a secondary pollutant that indirectly results from human activity. It occurs when organisms decompose organic matter naturally such as forests and oceans, but it is also found in products like farming and mining. Then there is the biogeochemical cycle. Biogeochemical cycles are the processes by which elements and compounds in Earth's biosphere are recycled (Metz et al., 2014). The carbon cycle, for example, refers to the exchange of carbon between the atmosphere, soil and living things (Carbon Cycle). The oxygen cycle refers to the process by which organic matter is broken down into organic compounds that plants can use as food while passing on reduced compounds like carbon dioxide (CO2) back into the environment (Oxygen Cycle).

- Biofuels

Biofuels are a major source of atmospheric carbon dioxide. These include ethanol, biodiesel and biogas. Biofuels can be used in cars, trucks, trains and airplanes. They are also used in power plants to generate electricity. Biofuels are of several types, and they can be used as a direct substitute for gasoline and diesel fuel. Using biofuels is environmentally friendly and reduces the carbon footprint by almost half. Biofuels, especially ethanol and biodiesel, are renewable fuels. They are derived from biological sources that can be grown, harvested and processed into useful energy products. These fuels are produced from biomass--primarily cellulose, sugar, vegetable oils and animal fats--that can be harvested sustainably without damaging the environment.

Biofuels can also be synthesized from different substances, such as lignocellulose (ST facility), plant oils and fats, animal fats, greases, liquid tallow and recycled cooking oil, in the form of gas or liquid fuels like biodiesel.

- Methane hydrates.

Methane hydrates are a solid form of methane that contain water molecules. They are found on land and at the bottom of the ocean floor in large deposits, called "methane hydrate reservoirs." Methane hydrates are currently a potential source of both methane and carbon dioxide (CO2). The release of these gases could have severe impacts on our climate (Chemat et al., 2012). The release of methane from permafrost soils or sub glacial sediments due to warming temperatures could accelerate global warming even more than current estimates suggest.

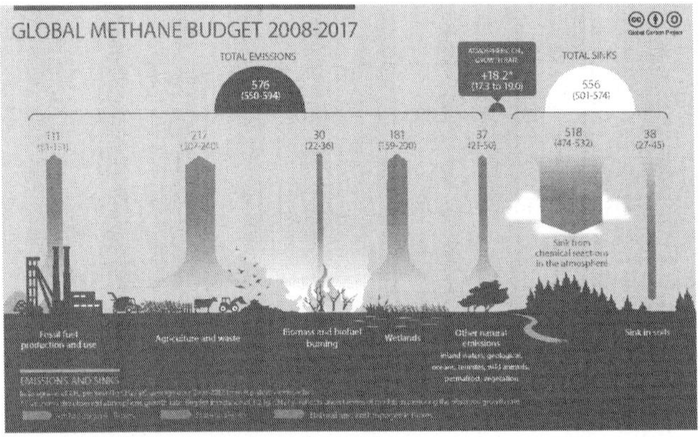

Figure 9.6: Primary sources of methane releases.

Source: By The Global Carbon Project - http://www.globalcarbonatlas.org/en/CH4-emissions / https://essd.copernicus.org/articles/12/1561/2020/, CC BY 4.0, https://commons.wikimedia.org/w/index.php?curid=93292720

Nonetheless, it is an important greenhouse gas that contributes to global warming. Methane hydrates are a gas store that have been known since the 1970's and this store can be tapped into by drilling or natural seepage. These are both proven methods of accessing methane (and as such CO2) from underground reservoirs, but they are not yet used on a commercial scale. Coastal areas in the northern hemisphere are rich in methane hydrates. These cold ocean areas, where the molecules of water and gas are trapped within the crystal structure of ice, harbor hydrate deposits that contain huge reserves of methane. One cubic foot of such a deposit contains over 100 times as much methane as all of the confirmed and estimated fossil fuel reserves on land and in deep waters. Methane hydrate resources are estimated to be the largest of all global gas reserves, and it is possible that they hold natural gas on a volume basis equal to all other fossil fuels in the world. The hydrate is a solid with the chemical formula $CH_4 \cdot 6H_2O$ and one of several compounds called clathrates. It is an amorphous form of natural gas consisting of a cage-like crystal lattice that encloses methane molecules within its structure. Methane is the second most abundant hydrocarbon in Earth's atmosphere, exceeded only by water vapor (Dubernet et al., 2010). Methane occurs either suspended as gas or deposits in ice crystals. As it releases energy when above -260 degrees Fahrenheit, it could be used as fuel in some forms.

- Cement production.

Cement production is one of the largest human sources of CO_2. The process uses a lot of energy and emits large amounts of carbon dioxide (CO_2). These emissions can be divided into two main categories: carbon dioxide (CO_2) and methane (CH_4). The rapid increase in the rate of cement production has contributed to a dramatic increase in carbon dioxide emissions over the past two decades. The release of CO_2 resulting from cement production in the USA and its use is responsible for approximately 9% of the global CO_2 inventory, making it second only to China on a per capita basis.

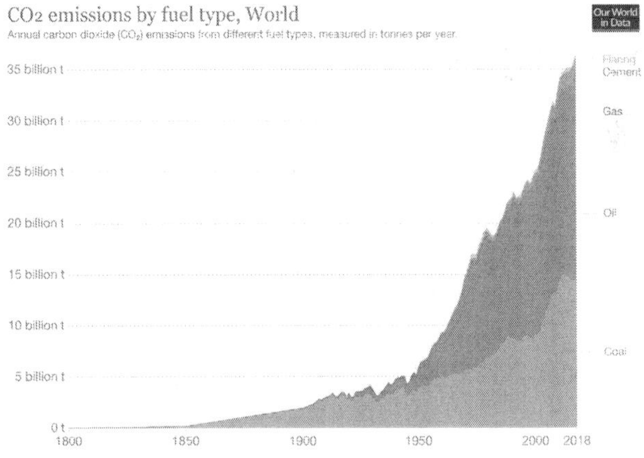

Figure 9.7: The making of cement reveals the production of gases that end up in the atmosphere.

Source: https://en.wikipedia.org/wiki/File:CO2_Emissions_by_source_%28Coal,_Oil,_Gas,_Cement,_Flaring%29_2018.png

- Methane gas from plants/swamps

Methane gas is produced by bacteria in the soil, by bacteria that live in the stomach and intestines of animals, and even by some plants (which is why you can smell rotting vegetation).

There are countless sources of gases in the atmosphere that contribute to global warming and climate change. Some come from natural processes like volcanic eruptions, while others come from anthropogenic sources such as fossil fuel burning or cattle farming. Either way, all these gases play a role in heating up our planet and could lead to significant consequences down the line if we don't start taking steps now to reduce them.

9.4 DISSOLUTION OF GASES IN SEA WATER

The dissolution of gases in sea water is described by Henry's law. This law states that the concentration C of a particular gas dissolved in water is proportional to the partial pressure P of that gas above the solution, where K is Henry's constant for that gas (Dufrêne et al., 2017). The constant K depends on temperature and salinity, both of which vary widely across the world.

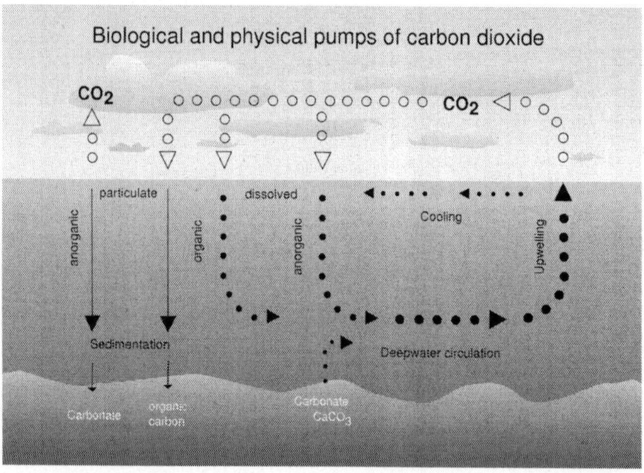

Figure 9.8: The impact and dissolution of gases into sea water also affect the atmosphere

Source: By derivative work: McSush (talk)CO2_pump_hg.png: Hannes Grobe 21:52, 12 August 2006 (UTC), Alfred Wegener Institute for Polar and Marine Research, Bremerhaven, Germany - CO2_pump_hg.png, CC BY-SA 2.5, https://commons.wikimedia.org/w/index.php?curid=4956173

The most important gas dissolving into sea water is oxygen. At high temperatures, water can dissolve 15 times more atmospheric CO than at its freezing point. Gas hydrates (GH) are the natural occurrence of gases in a solid state, in a crystal structure within an ice-like matrix. The major gases found in hydrates are methane and carbon dioxide. Gas hydrates have become of interest as a potential source for hydrogen for fuel cells and as an energy storage medium because of their high content of hydrogen combined with the low environmental impact of this application. In the atmosphere, gaseous carbon dioxide dissolves and reacts with water in order to form a weak acid (carbonic acid), which can be detected by its sharp taste.

9.5 PHYSICAL PROPERTIES OF GASES IN THE ATMOSPHERE

- Motion

Gases are not static. They're always in motion, and they move because of convection (heat rising), turbulence (random motion), friction, or pressure. This is true even at the Earth's surface: when you wave your hand around, you can feel a slight breeze caused by convection and turbulence. In the atmosphere, gases move upward through vertical motions called convection (Ebbesen, 2016). There are several different kinds of vertical motions such as advection and diffusion that can cause gas molecules to move from one place to another without being affected by gravity; these are called horizontal motions because they occur parallel to the ground instead of vertically up toward space.

- Pressure

When you blow up a balloon, what happens? The air inside the balloon exerts more pressure than the air outside. We call this pressure exerted by gases atmospheric pressure. The force of gas molecules pushing against each other is called an interaction force and is also called an intermolecular force. These forces are stronger for smaller molecules (like helium) than for larger ones (like nitrogen). In addition, these interactions between molecules depend on temperature: as temperature increases and energy reaches its maximum level, more vibrations occur with more kinetic energy resulting from higher temperatures; so these vibrations get closer together until they reach the point where they can be considered touching one another at all times – regardless of how far apart they were before! This causes them to repel each other because there are too many collisions occuring right next door instead of further away like before when there weren't any particular "directions" which could take place between them all over again every single second throughout their entire lives together on earth

- Relation between temperature and pressure

The gas law relationship shows that temperature and pressure are related. Pressure is the force exerted by a gas per unit area of its surface. The units of pressure are pascals (Pa), where 1 Pa = 1 N/m2. The relationship between these variables can be represented by an equation that states that if you have two containers with different volumes and equal amounts of gas, then their pressures will be equal (assuming no outside influences). Here's how to use this equation: Say we have two containers with the same amount of oxygen in them at different temperatures (one hot, one cold) (Gentile et al., 2014). The hot container has less room for expansion so it will feel tighter than the cold one which has more room for expansion; thus, it feels looser.

- Molar mass

Gases may be measured in a number of ways. The most common method is to measure their volume and pressure. Using these measurements, it is possible to calculate the density of a gas using the ideal gas law: PV=nRT where P is pressure, n is moles (or molecules), V is volume, and R is an ideal gas constant. The molar mass can then be calculated by dividing the mass of each component by its molecular weight (MW).

- Concentration.

The concentration of gases in the atmosphere is not uniform. Atmospheric gases are not evenly distributed throughout the atmosphere, nor are they uniformly distributed on earth.

Air Properties Definitions

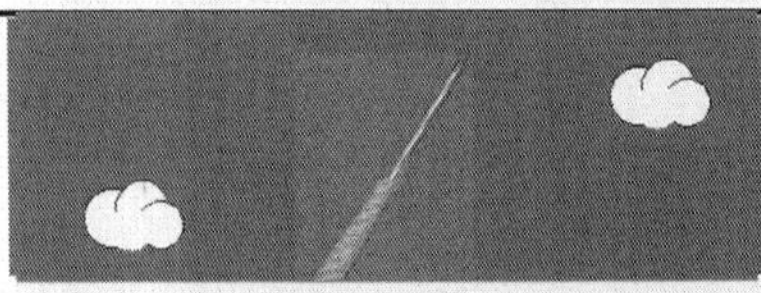

Air is a Gas. 78% Nitrogen, 21% Oxygen, traces H_2O, CO_2, Ar,..

Property	Dimensions	Value (SLS*)	
Mass, Volume		Metric	English
Density (r)	mass/volume	1.229 kg/m^3	.00237 slug/ft^3
Specific Volume (v)	volume/mass	.814 m^3/kg	422 ft^3 /slug
Pressure (p)	force/area	101.3 kN/m^2	14.7 lb/in^2
Temperature (T)	degrees	15 °C	59 °F
Viscosity (mu)	force-time/area	1.73 x 10 N-s/m^2	3.62 x 10^{-7} lb-s/ft^2

* Sea Level Static (Standard Day)

Figure 9.9: A NASA generated air property definitions.

Source: Image by NASA

- Density

Density is a measure of the mass of a substance per unit volume. It can be expressed in units of grams per cubic centimeter (g/cm3) or kilograms per cubic meter (kg/m3). The density of air at sea level is approximately 1.2 kg/m3 and decreases with increasing altitude. Density is related to the molar mass of a substance and its mass:

Where M = molar mass, R = ideal gas constant, T = temperature in kelvin, p = pressure. Density is a physical property of gases that are compressed by gravity. The density of a gas can be calculated using the formula: density = (mass of gas) / (volume).

- Inertia

Inertia is a property of all matter. It is the tendency of an object to resist changes in its motion. For example, if you accelerate a box from rest to 10 meters per second in one second, then it has an inertia (or mass) of 10 kg m/s2. The principle of inertia is one of the key physical properties of gases in the atmosphere that can cause meteorological and oceanographic phenomena, such as turbulence and cyclogenesis (Haidekker & Theodorakis, 2010).

To understand inertia, imagine that you are standing on a skateboard and start rolling down a hill at constant speed. If you push on your right foot, it will cause the left side of the skateboard to move forward faster than the right side—the left end of your skateboard will be going faster than before while they both continue moving at constant speed downhill; this is called being underdamped or overdamped depending on how quickly they respond relative to each other.

Inertia is a physical property of gases in the atmosphere. Essentially, inertia as a property is only useful for gases when it comes to interacting with other forces. When gas molecules come into contact with one another, they tend to bounce off and move out of each other's way from an opposite direction. Because of this, gases are considered to have an extremely low frequency of interactions with each other and thus extremely low inertia.

Inertia plays an important role in the Earth's atmosphere, especially as a force that drives our weather. The inertia in the atmosphere causes air to move up and down when temperature changes occur, resulting in friction between layers of air. The friction causes kinetic energy to be released as heat and sound. The rate of heat loss by convection depends on the temperature difference between layers and how fast convection occurs. Increased humidity, which results from increased evaporation associated with high temperatures, slows down convection by increasing its resistance. When there is a large temperature drop within a layer of air, air can cool rapidly enough to reach its dew point, leading to condensation and precipitation (clouds).

- Viscosity

Viscosity is a measure of how viscous or thick a fluid is, and it's caused by friction between the molecules within it. Viscosity can be thought of as a fluid's resistance to flow: if you put a drop of water on your finger and try to move it around in circles, you'll notice that it flows easily. If you try the same thing with honey, however, there will be more resistance—it won't move so freely. The difference in viscosity between these two substances is caused by their different molecular structures: while both have billions of tiny particles moving through them at once, those particles are arranged differently in each substance (Hasanuzzaman et al., 2013). Honey has an irregular structure due to its high sugar content; water has an orderly crystalline structure (like in ice). This means that honey resists moving more than water does because its molecules aren't packed tightly enough together for them all to flow smoothly past each other without any loss of energy.

Viscosity is a physical property of gases in the atmosphere that influences the flow and dispersion of pollutants. Viscosity is determined by temperature, humidity, pressure and molecular weight of oil. Viscosity increases with increasing molecular weight and decreasing temperature. The viscosity of a gas can be used to help identify the gas type, or to find the relative concentration in an atmosphere. Viscosity is defined as the resistance to flow through a fluid.

- Compressibility

The compressibility of a gas is the ability of the gas to be compressed. It depends on the molecular weight, temperature, pressure and density. The compressibility of a gas is related to its density and is an important factor in determining the behavior of air masses. As you increase pressure on a gas its volume decreases, decreasing its density. The capacity of air to compress tells us how much the pressure goes up when the volume goes down at a given constant temperature.

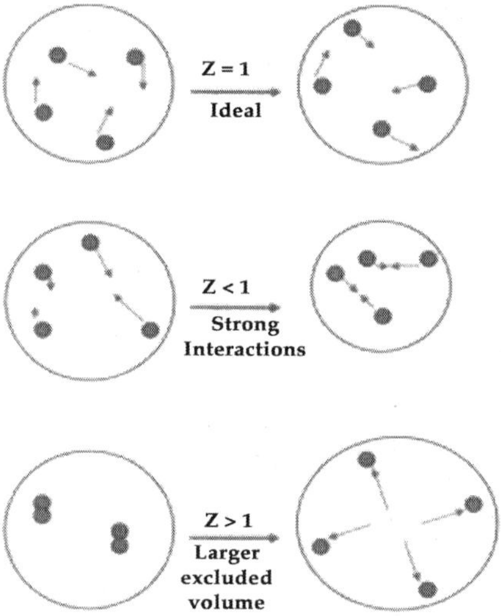

Figure 9.10: A test on the compressibility of gases.

Source: By BlyumJ - Own work, CC BY-SA 4.0, https://commons.wikimedia.org/w/index.php?curid=64937414

It has units of pressure per unit volume (Pa m3). Compressibility can be considered a mechanical property, since it depends on the elasticity of atoms or molecules and elasticity is defined as a mechanical property. Therefore, based on this definition it would be correct to consider compressibility as a mechanical property of gases in the atmosphere. In other words, we can say that compressibility depends on temperature more than pressure.

- Thermal expansion

When a gas is heated, it expands. The more it heats up, the more it expands. This is called thermal expansion and it's the opposite of thermal contraction, which happens when something cools down (like water). The law of thermal expansion states that if you heat a gas just a tiny bit at constant pressure and then measure its volume again at the same constant pressure (and then repeat this process many times), you'll find that its volume increases by about 0.5% for every 1°C increase in temperature above absolute zero (Mendoza & Núñez, 2009).

This makes sense: when something is hot, its molecules have more energy than when they're cold; so they move around faster and spread out more widely from each other as space becomes available to them. The reverse happens with coolingbecause when something gets cooler all its molecules settle down into lower levels of energy so there isn't much room left over for any new ones to move around freely without bumping into one another. In fact, because the average speed of particles inside solids stays pretty much fixed once their temperatures are reached, whereas those inside liquids vary greatly depending on whether they've been heating up or cooling down!

Thermal expansion is a physical property of gases in the atmosphere due to which their volume increases with increase in temperature. Thermal expansion is a physical property of gases at constant pressure. When a gas expands, the individual molecules are farther apart, allowing more empty space between them. This means that at a given temperature, there is more space available for outside influences to fill up.

Knowing the physical properties of gases in the atmosphere is important for understanding the kinds of weather and climate we can experience. For example, density, inertia and viscosity are all important factors in determining how air moves through mountains (Mansour et al., 2010). Air near sea level is less dense than air at higher altitudes because it contains fewer particles per unit volume than higher-altitude air. This means that lower-density air will rise more easily, making it easier for wind to move over mountain ranges rather than around they (think about how an inflated balloon moves when you blow on one side).

Viscosity also plays an important role when thinking about how air moves around mountains—viscous forces tend to resist movement through a substance like water or smoke and make flow slower than expected. In this case, viscosity slows down the movement of air across mountain ranges so that winds don't just blow straight over top of them but instead curve around their sides before heading back down into lower altitudes again (think about blowing bubbles through a straw). Lastly, thermal expansion causes gases like nitrogen dioxide (NO_2), carbon dioxide (CO_2) and ozone (O_3) to expand when they are heated up—but not so much when they're cooled down!

CHAPTER 10

LIQUID, SOLID, AND OTHER MATERIALS

CONTENTS

The structural, mechanical, and dynamical properties of foam structures are thoroughly investigated. On the one hand, liquid or semi-solid forms are explored, but solid foams are also explained. Metallic, ceramic, and biological foams are all subclasses of each other in terms of cell structure. Solid foams can be employed for their enormous surface area, acoustic, and thermal qualities, in addition to their light mechanical properties. In current biomaterials, ceramic or organo-ceramic foams are utilized to construct bone scaffolds, whilst hierarchically micro-and nanoporous structures are employed in chemistry to regulate catalytic processes.

Natural and synthetic foam forms and characteristics will play a growing role in the fabrication of materials in the future, from food packaging to vehicle chassis, which bode well for future foam research and development. Atoms in a newly found kind of matter may be either solid or liquid. Previously, atoms in physical materials were assumed to exist in one of three states: solid, liquid, or gas. According to studies, when exposed to severe temperatures, certain elements can display both solid and liquid qualities (Marian, 2012). High pressure and temperature lead the majority of potassium atoms to form an extraordinarily stable lattice structure. In contrast, the second pair of potassium atoms in the structure is in a fluid condition. Researchers predict that if given the right conditions, a variety of additional elements, including sodium and bismuth, will be able to flourish in the newly discovered form.

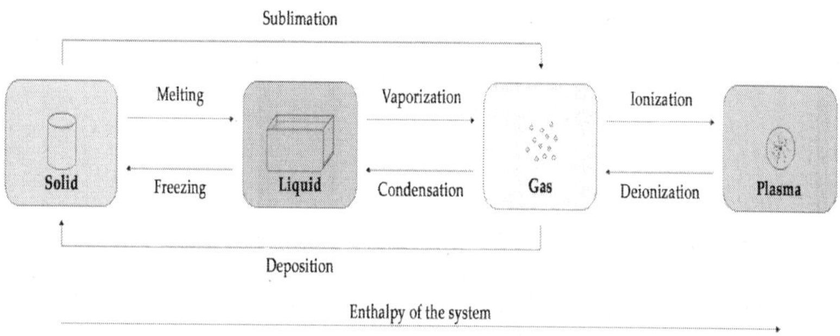

Figure 10.1: A representation of the states of matter.

Source: By ElfQrin - Own work, CC BY-SA 4.0, https://commons.wikimedia.org/w/index.php?curid=15779425

A liquid's volume is fixed, but it can flow and take on the shape of its container. This is because the gap between particles is large enough for them to pass through one another. Here are some liquid examples: In descending order of importance: 5. The world has a lot of water (a liquid metal) The sixth element is oil. Milk is seventh on the list. Eighthly, acetone It's the eighth alcoholic drink. Ten cups of coffee Gases are used as visual aids. Argon may exist in all three states at the same time (Matranga & Corsi, 2012). Temperature and pressure may have a considerable influence on the characteristics of a material. Because gases have no definite volume or structure, they can fill containers of any size.

Gaseous particles are widely different from liquids and solids. Excellent examples include carbon dioxide, methane, nitrous oxide, and hydrogen-air. A few instances lack a recognizable structure and volume. It can stretch to fill a container. This particle is ionized (has an electric charge), making it distinct from other types of particles. Lightning is a great example of plasma in action.

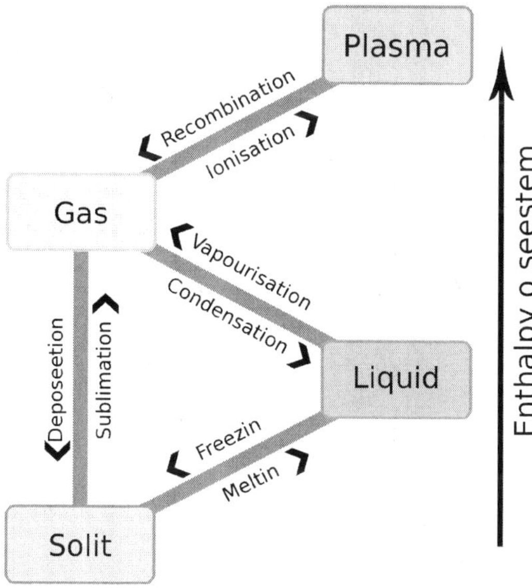

Figure 10.2: States of matter.

Source: https://en.wikipedia.org/wiki/File:Phase_change_-_en.svg

Although the most prevalent states of matter are solids, liquids, gases, and plasma, scientists are aware of an unlimited number of different forms of matter. Here are a few examples of how this may be accomplished: "Liquid crystals" are crystals that exist in liquid form. A viscous liquid-like substance is an example of a **superfluid**. The Bose-Einstein condensate in this case: A Bose-Einstein condensate occurs when particles in a super cold gas can no longer interact with one another. In a color-glass condensate: It is believed that this material exists in the nuclei of atoms flying at near-light speed (Lovley, 2003). A material that absorbs but does not emit light is referred to as "dark matter."

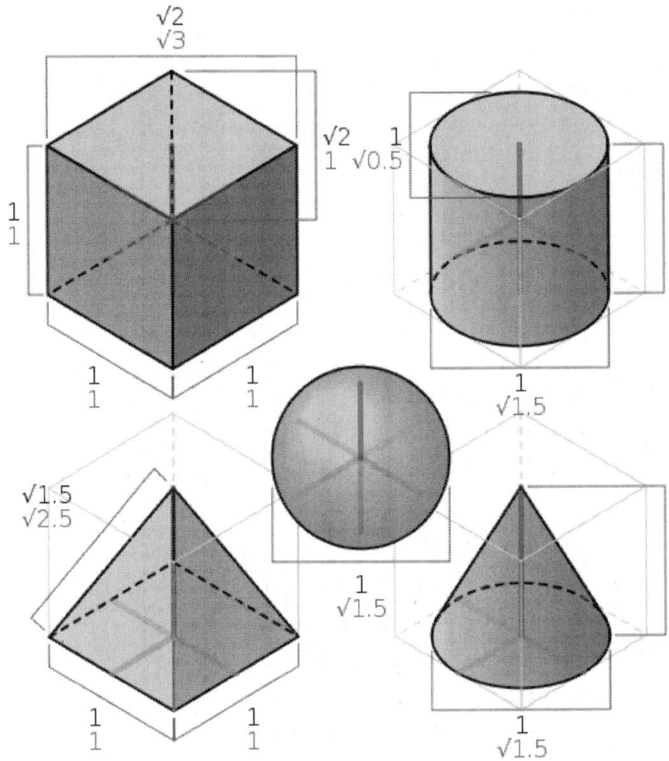

Figure 10.3: Solid shapes.

Source: https://en.m.wikipedia.org/wiki/File:3D_shapes_in_isometric_projection.svg

Temperature and pressure fluctuations cause a change in the state of matter. The following are the most typical phase transitions: The act of turning a liquid into a solid is known as "freezing." Deposition: the change of a gas into a solid is known as a deposition. Melting is the transformation of a solid into a liquid. Condensation of gas results in the formation of a liquid. When a liquid sublimate into a gas, this is referred to as vaporization. Plasma is converted into a gas during recombination, also known as deionization. Ionization happens when a gas is converted into plasma. Depending on temperature and pressure, the matter may experience phase transitions: solids can melt into liquids, solids can sublimate into gases, and liquids can evaporate into gases. Solids form from liquids, while liquids form from solids.

When temperature and pressure fluctuate, particles might migrate further apart. As a result, solids become liquids and liquids become gases. This indicates that a solid may become a gas without going through the liquid phase, and vice versa for a gas. For the first time, researchers have created materials that can permanently change from solid to liquid or vice versa when exposed to light at room temperature and remain in a new phase long after the light is turned off. Light may also be utilized to build liquid patterns inside solid materials or solid patterns within liquid materials, resulting in stable half-solid, half-liquid solids (Lischka et al., 2011). Novel materials may be employed in several methods other than 3-D printing, molding, and on-demand recycling. All that we see and experience around us is made up of matter. The term "matter" can be bewildering due to its numerous connotations. The question "What's wrong?" is frequently asked. "It makes no difference," for example. The matter is described by scientists as "everything that occupies space and has mass." Micro particles are the fundamental constituents of matter.

Because atoms and molecules are both made up of atoms, this might apply to either. Atoms are similar to LEGO blocks. They are the smallest unit that can be destroyed without resorting to drastic means, unlike LEGO bricks or atoms in the Large Hadron Collider. The atomic and molecular properties of matter define its structure or state. Your mouse and keyboard are generally on the desk next to you while seated in your chair. A solid thing is difficult to compress (squish) because it preserves its form (squash). Most solids have densely packed particles. The particles aren't moving and can't travel through each other, but they do vibrate somewhat. Ice is formed when water crystallizes. It is possible to remove ice from a container while keeping its original shape. Ice's molecules are more loosely packed than those of liquid water, making them distinct from other solids. This helps to explain why ice floats in water.

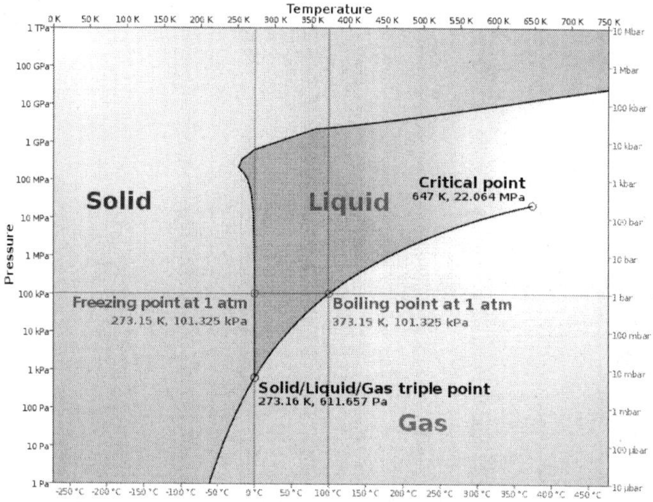

Figure 10.4: Freezing, melting and boiling points of water.

Source: By author of the original work: Cmglee - Own work, CC BY-SA 3.0, https://commons.wikimedia.org/w/index.php?curid=34865054

You may detect if something is liquid or not by simply checking if it fits into the new container when moved from one to another by pouring, for example). When you pour water into another glass, it's surprising how perfectly it fits. Water that has been spilled can spread swiftly. Because it is not limited, it produces a big puddle.

Liquids may flow and pass one other more freely when there are fewer particles packed together. Consider how difficult it is to compress water into a small container, despite the idea that compressing liquids is simpler than compressing solids. Atoms and molecules spread more easily in gases than in solids and liquids. They vibrate and move swiftly. Any container may be filled with gas, but if the container is not correctly sealed, the gas will leak out. A gas's compressibility is significantly larger than that of a liquid or solid. Consider a diving tank with a 3-liter cylinder storing 600 liters of compressed gas (Huang & Zou, 2010). The air you are breathing contains several elements, including oxygen and nitrogen. Water in its gaseous state is referred to as "water vapor." Unlike ice or water, water vapor is undetectable. We expel water vapor as we exhale. The water vapor we exhale is invisible, but we can see it condensing if we hold our glasses or smartphone to our lips (becoming liquid).

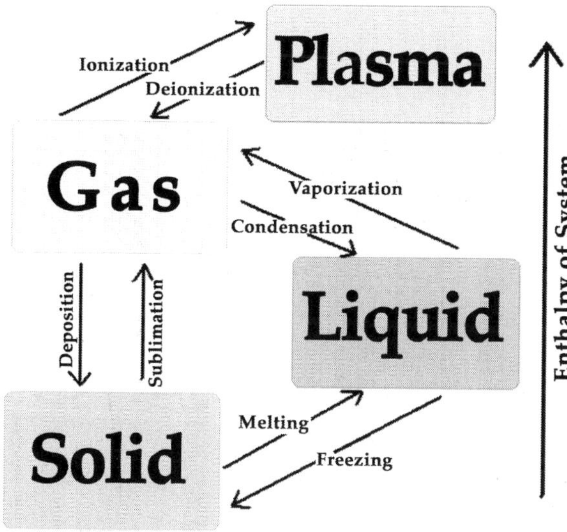

Figure 10.5: State of matter processes.

Source: https://upload.wikimedia.org/wikipedia/commons/9/97/Phase_change_-pn.png

Researchers have discovered previously undiscovered chemicals in a variety of forms as a result of unexpected findings. Plasma is used in fluorescent light bulbs and plasma displays because it is a natural phenomenon that occurs in lightning and is produced by people. Bose-Einstein condensate is a state of matter that can only be found at extremely low temperatures.

Solids are one of the four fundamental states of matter, along with liquids, gases, and plasma. A solid's molecules are densely packed and have the lowest kinetic energy of any material. The structural stiffness and resistance to external pressures on a solid's surface characterize it. In contrast to a liquid, a solid does not flow or expand to fill its container to its full capacity. When it comes to solids, atoms can be arranged in either a regular geometric lattice (as in crystalline solids such as metals and ice) or randomly (as in conventional window glass)). Solids cannot be squeezed at low pressures due to the loose packing of molecules in a gas, whereas gases may be crushed at low pressures.

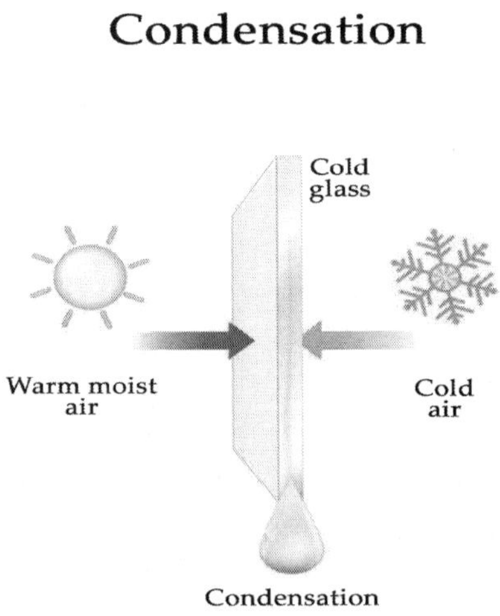

Figure 10.6: The condensation process.

Source: https://media.istockphoto.com/vectors/condensation-on-a-window-vector-id1292579909?k=20&m=1292579909&s=612x612&w=0&h=D7Ky-zXPFncmYANsuKePEg484UDdtvu8tfazOPNkjg0=

Solid-state physics, which works with both liquids and gases, is one of the most significant subfields in condensed matter physics (which also includes liquids). Materials science is concerned with the physical and chemical properties of solids. New materials are created in solid-state chemistry, and their chemical composition and identification are explored. Solids are created by arranging atoms, molecules, and ions in either a symmetrical or asymmetrical form. There are two kinds of crystals: crystallized crystals and non-crystallized crystals. The large-scale structure may survive without interruption in some contexts, such instance with diamonds, where each diamond is a distinct crystal. Many huge solids that can be seen and touched are made up of crystallites, which can be as tiny as a few nanometers or as massive as a few meters. Polycrystalline substances are composed of numerous crystals. Except for a few, polycrystalline materials include practically all common metals and ceramics (Lisio et al., 2019). Other atoms are not organized in a long-range pattern in few molecules. This type of material includes amorphous solids such as polystyrene and glass. The material and conditions under which a solid was formed decide whether it is crystalline or amorphous.

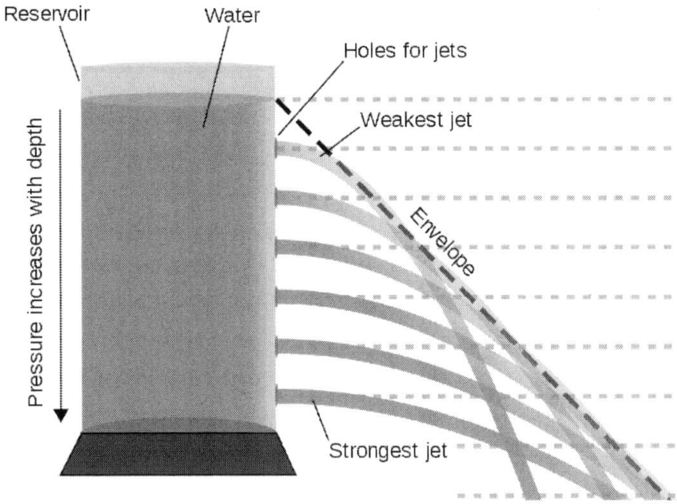

Figure 10.7: Water pressure experiment.

Source: By Matt Cook - Own work, CC BY-SA 4.0, https://commons.wikimedia.org/w/index.php?curid=45705236

Crystalline materials are more prevalent in slowly cooled materials, whereas amorphous solids are more abundant in immediately frozen materials. The structure of a crystalline solid is also determined by the material from which it was made, as well as how it was generated. Aside from ice cubes and coins, which are formed of the same fundamental molecules throughout, many other items have a wide range of chemical compositions. A typical rock, for example, is composed of several minerals and mineraloids that cannot be recognized chemically. The major components of wood, which is a naturally occurring organic substance, are cellulose fibers and lignin. In the realm of materials science, composites may be constructed from a range of materials with the appropriate qualities.

There are several varieties of solids. Atoms in a solid are susceptible to a variety of forces. A crystal of sodium chloride, for example, is held together by ionic sodium and chlorine bonds. Covalent bonds are produced by sharing electrons between diamond or silicon atoms. Metallic bonding is the process through which metals share electrons. Van der Waals forces, which occur from the polarization of each molecule's electrical charge cloud, hold some substances together, including the vast majority of organic molecules. The differences between solids are caused by differences in bonding. The density and strength of metals make them excellent for use in electrical and thermal applications. Moreover, half of the elements in the periodic table along a diagonal line from boron to polonium are metals. Alloys are mixtures of two or more elements, with metal constituting the majority (Laganà et al., 2019). Metal use has been a long-standing human habit that dates back to the beginning of recorded history. Metals are widely used in the construction of buildings and other structures, in addition to the great majority of cars, numerous appliances and equipment, pipelines, road signs, and railway tracks. Iron and aluminum are the most prevalent structural metals. These metals, which are also the most abundant, dominate the Earth's crust. Steel, as an alloy, contains up to 2.1 percent carbon and is hence tougher than pure iron, which is most usually utilized.

Metals are used in electronic equipment and long-distance transmission of electrical current because they are good conductors of electricity. As a result, metal cables are used for power distribution in electrical networks. Copper is widely used in domestic electrical devices due to its high conductivity and ease of machining. Because of their high heat conductivity, metals are suitable for stovetop cooking equipment. The study of metallic elements and their alloys is heavily influenced by solid-state chemistry, physics, materials science, and engineering. A high number of delocalized, shared electrons make up the "metallic bonding" that keeps metallic solids together. Positive ions are created when metal atoms rapidly lose their outermost "valence" electrons. Electrostatic interactions between the ions and the electron cloud hold the solid together. Metals have a lot of free electrons, which explains their great electrical and thermal conductivity. Because the free electrons hinder the transmission of visible light, metals become opaque, glossy, and shiny. Complex metal property models can account for electrons that have been delocalized. Because most metals have a crystalline structure, the ions are usually grouped in a periodic lattice. The practically free electron model is the simplest of many theoretical models for dealing with the ion core potential.

Figure 10.8: Ice cubes.

Source: By Darren Hester - Openphoto.net, CC BY-SA 2.5, https://commons.wikimedia.org/w/index.php?curid=1452658

10.1 MINERALS

Minerals are naturally occurring solids that develop under high pressure as a result of a variety of processes. Mineralized compounds must be crystalline with uniform physical characteristics. Minerals can range from basic salts and pure elements to hundreds of different silicates. A rock sample contains an unknown number of minerals and/or mineraloids. Quartz (crystalline SiO2), feldspar, mica, chlorite, kaolin, calcite, olivine, augite, hornblende, magnetite, hematite, and a few more minerals make up the vast bulk of Earth's crustal rocks. While quartz, mica, and feldspar are common minerals, others have only been identified in a few places across the world (Kalra et al., 2016). The most prevalent mineral class is silicates, which are largely composed of silicon and oxygen, as well as ions of aluminum, magnesium, iron, calcium, and other metals (95 percent of rocks are silicates). Ceramics Inorganic chemicals used to manufacture ceramic solids are primarily made up of chemical element oxides.

They remain chemically inert when exposed to acidic or alkaline environments. Inorganic compounds that do not oxidize include nitrides, borides, and carbides. Clay minerals like kaolinite are typical ceramic raw ingredients, although aluminum oxide (alumina) is a more recent addition (alumina). Silicon carbide and tungsten carbide are both relatively new ceramics with significant technological complexity. Abrasion resistance is an important consideration when choosing materials for mining crusher wear plates, which is where these two come in. Light scattering centers in most ceramics, such as alumina and its derivatives, have wavelengths that match the visible spectrum of light. It looks to be transparent, yet this is not the case. It is now feasible to make polycrystalline transparent ceramics such as transparent alumina and alumina compounds for high-power laser applications using nanoscale (e.g., Sol-gel) technologies. Medical, electrical, and electronic sectors also use advanced ceramics.

Ceramic engineering is concerned with understanding and developing solid-state ceramic materials, components, and devices. This is possible by heating or precipitating chemical solutions at lower temperatures. Chemical synthesis, compound synthesis, component synthesis, and component analysis are all part of this broad idea. Fibrillation, fracture toughness, compression strength, shear, and tension weakness are all mechanical features of ceramic materials. The ability to withstand a static load can improve the tensile strength of weak materials (Kleber & Johnson, 2010). Fracture toughness refers to a material's ability to survive fracture through crack development and propagation in the presence of intrinsic microstructural faults, whereas toughness refers to how much energy a material can accept before mechanical failure. A material with a high fracture toughness rating will most likely fracture ductilely, according to the fundamental principles of fracture mechanics. Because most ceramic and glass-ceramic materials have low k values, brittle fractures are common. Organic solids include wood, paraffin wax, naphthalene, and a variety of polymers and plastics.

Figure 10.9: Precipitation: Copper from a wire is displaced by silver from a silver nitrate solution it is dipped into, and metallic silver crystals precipitate onto the copper.

Source: By Toby Hudson - Own work, CC BY-SA 3.0 au, https://commons.wikimedia.org/w/index.php?curid=24987172

10.2 WOOD

Wood is a natural organic substance since the bulk of its structure is made up of cellulose fibers and lignin. Tensile strength is provided by the lignin matrix, which is also compressible. Wood was an important component when people first began building shelters and boats. "Construction-grade wood," "lumber," and "timber" are all phrases that are sometimes used interchangeably. Wood is used in building as both a structural component and a concrete mold. Refined pulp is heavily dependent on wood-based goods and is used in both packaging (e.g., cardboard) and paper. Before burning lignin, Kraft or sulfite chemicals are employed to destroy its chemical bonds (Karsenti, 2008). Polymers Carbon's capacity to generate particular compounds with molecules capable of covalently joining and creating a chain or network is crucial in organic chemistry. Polymerization is the process of creating polymer chains or networks using a monomer as the starting component. Biopolymers and industrial polymers are the two forms of polymers found in nature.

Substituents and functional groups can influence the chemical and physical properties of organic compounds, including solubility and chemical reactivity, as well as physical properties such as hardness and density, mechanical or tensile strength, abrasion, heat resistance, transparency, and color. Because of these properties, proteins can preferentially take on a physiologically active form over others. For ages, people throughout the world have used natural organic polymers such as waxes and shellac, which are thermoplastic polymers. Natural rubber and cellulose, a plant polymer, were frequently used in the early nineteenth century to boost the tensile strength of natural fibers and ropes. Polymers, which are the building blocks of resins, are essential for the manufacture of plastics. Polymers and other additives are added to a resin throughout the production process before it is molded into its final shape. Polyethylene, polypropylene, polyvinyl chloride, polystyrene, nylons, polyesters, acrylics, polyurethane, and polycarbonates are the oldest and most often used polymers. Plastics are commonly used to refer to commodity, specialized, and engineered plastics. Composite-derived materials in composite materials, ceramic is frequently one of several macroscopic phases.

10.3 SEMICONDUCTORS

Semiconductors sit between metals and nonmetals in terms of electrical conductivity and resistance. They may be found in the periodic table diagonally descending from boron. These symbols differentiate between electrical conductors (or metals) and insulators (to the left of them). Radios, computers, and telephones are examples of semiconductor-based devices that underpin the current electronic industry. Semiconductors are electronic elements that include semiconductor devices such as transistors, diodes, and integrated circuits. A huge semiconductor device that transforms light directly into electrical energy is a solar photovoltaic panel. A metallic conductor is termed to be a "conductor" if an electron or positively charged "hole" in a material's electronic band structure transports current. Silicon, germanium, and gallium arsenide are the most prevalent semiconductor materials. Nanomaterials: When common materials are reduced to nanoscale dimensions, a wide range of characteristics emerge. Yellow gold nanoparticles have a red tinge, as do gray silicon nanoparticles; gold nanoparticles melt at a lower temperature than gold slabs (300°C for a 2.5 nm size), and metal. Nano wiring is stronger than bulk metal (Krylov & Gill, 2013). Nanoparticles are useful for a wide range of energy-related applications due to their large surface area. Platinum metals, for example, can be useful in enhancing automobile gasoline catalysts and proton exchange membrane (PEM) fuel cells.

Ceramic oxides (or cermets) containing lanthanum, cerium, manganese, and nickel are being developed for use in solid oxide fuel cells (SOFC). Lithium, lithium-titanate, and tantalum nanoparticles are utilized in lithium-ion batteries. During the expansion/contraction cycle, silicon nanoparticles substantially boost the storage capacity of lithium-ion batteries. Because silicon nanowires do not degenerate after cycling, they might be used in batteries with significantly longer storage periods. Silicon nanoparticles are also being used to create new types of solar cells. The power output of photovoltaic (solar) cells can be enhanced by up to 60% by covering the polycrystalline silicon substrate with a thin layer of silicon quantum dots and fluorescing incoming light before capture. To boost radiation absorption, the surface area of nanoparticles and thin films must be increased. Mechanical characteristics of biopolymers and other natural composites are outstanding. Scientists are encouraged to create new materials as a result of these complicated designs. It is one of a kind since it has a hierarchical structure, is multipurpose, and can self-heal. Self-organization is essential in all biological systems, from the tiniest molecules to the most complex biological systems. As a result, self-assembly, a novel technology for producing high-performance biomaterial chemicals, is gaining attention. What components makeup the odor? What are the characteristics of the odor? Physical qualities of elements and compounds, such as hardness, porosity, index of refraction, and others, provide unmistakable proof of their chemical makeup. This section discusses the physical characteristics of solids. Mechanical Mechanics, among other things, describes a material's strength and deformation resistance. Steel beams, for example, are employed in construction because of their great strength, which prevents them from breaking or bending under severe loads (Kharat et al., 2017. Mechanical parameters to consider include tensile and compressive strength, compressive tensile and shear strength, fracture toughness, ductility, and indentation hardness. It is the study of how solid matter responds to external pressures and temperature changes.

Solids, unlike fluids, do not display macroscopic flow. Any deviation from the original shape is referred to as deformation. The difference between an object's new and prior sizes is defined as stress. Almost all solid materials obey Hooke's rule, which states that if the applied stress is low enough, strain is exactly proportional to it. Young's modulus is a term used to describe how elastic something is (or modulus). This region of deformation is known as the linearly elastic area. Three models may be used to demonstrate how a solid response to stress: When a substance is elastic, it returns to its normal shape once a strain is removed. This category includes materials that are both elastic and dampening. When the applied tension is relaxed, the damping effects must be overcome, resulting in the release of energy as heat within the material. As a result, the stress-strain curve produces a hysteresis loop. Because of this, the mechanical reaction changes over time.

Figure 10.10: Plastic bottles.

Source: By Amraepowell - Own work, CC BY-SA 3.0, https://commons.wikimedia.org/w/index.php?curid=19133754

- Elasticity - The material responds elastically when the applied stress is less than the yield value. When the stress surpasses the yield stress, the material does not **return** to its original shape. That is, there is irreversible plastic deformation following a permanent yield (or viscous flow). Many materials deteriorate at high temperatures. High-temperature refractory materials find usage in a variety of industries and applications. Glass ceramics have proven to be an outstanding countertop cooking material due to their superior mechanical properties and capacity to resist temperature changes of up to 1000°C. Thermal shocks must be able to be absorbed by materials used in the design of airplane and spacecraft exteriors (Kawser Hossain et al., 2016). Polymer/ceramic/metal composite materials, as well as fiber-reinforced polymers, are being researched for this purpose. Atoms can wander about in the lattice structure when heated sufficiently, although they prefer to stay near to the nucleus. The spectrum of lattice vibrations in a crystalline or glassy network is the foundation of solid kinetic theory. Because this movement occurs at the atomic level, spectroscopy is the only method for detecting and characterizing it. Thermal conductivity is a property of a substance that indicates its ability to conduct heat as one of its thermal characteristics. The ability of a substance to store energy in the form of heat (or thermal lattice vibrations) in the form of heat is referred to as heat capacity.

10.4 ELECTRICAL PROPERTIES

Electrical properties include conductivity, resistance, impedance, and capacitance. Glass and ceramics, both electrical insulators, compete with metals and alloys, both electrical conductors. There are several semiconductors in this region. In metals, electrons are responsible for conductivity, but in semiconductors, electrons and holes both contribute to current flow. Ions, on the other hand, keep the current flowing in ionic conductors. At low temperatures, several metals and metal alloys, as well as some heavily doped semiconductors and ceramics, exhibit superconductivity.

The electrical resistivity of most electrical (metallic) conductors decreases with decreasing temperature, although it remains finite. Superconductors, on the other hand, exhibit little resistance when cooled below their critical temperature. In the absence of a power supply, an electric current traveling around a loop of superconducting wire can exist eternally. A dielectric is a substance that acts as a barrier to the flow of electricity. This substance is also known as an electrical insulator. In capacitors, an applied electric field can be focused inside a dielectric, such as plastic (Antoni et al., 2015). When two closely spaced conductors (referred to as "plates") come together to create a capacitor, energy is stored in the electric field between them Capacitors feature two plates with opposite polarities of charge when a voltage is applied to them. To discriminate between high-frequency and low-frequency signals, capacitors can be employed in electrical circuits as well as electronic filters. Electro-mechanical It is the ability of a crystal to produce a voltage in response to mechanical stress that is known as piezoelectricity.

When piezoelectric crystals are subjected to an externally applied voltage, they can change shape. Polymer materials that behave as electrets include rubber, wool, hair, wood fiber, and silk. The piezoelectric sensitivity of polyvinylidene fluoride (PVDF) is several times larger than that of quartz (crystalline sio2) (crystalline sio2). Its low deformation can benefit high-voltage sources, speakers, lasers, chemical, biological, and acousto-optic sensors, and transducers (less than 0.1 percent). Optical materials such as glass and metal may either transmit or reflect light. The three main classes of materials to be understood are transparent, translucent, and opaque. Plastics and other materials with some transparency make certain that the light beams may move freely across the space. Two such examples are pure water and crystal-clear glass. It's feasible to permit specific light beams through transparent materials. Contaminated goods include water, oil, and even filthy glass. Light cannot penetrate opaque materials. Examples include wood, iron, and plastic. Many materials let some wavelengths pass through while blocking others. UV photons that because sunburn is invisible to the human eye, but most of them are blocked by window glass. This property is shared by frequency-selective optical filters, which can change the color of incident light.

Figure 10.11: Electricity power.

Source: By I, Dingy, CC BY-SA 3.0, https://commons.wikimedia.org/w/index.php?curid=2471716

CHAPTER 11

FUELS, ORGANIC CHEMICALS, AND POLYMERS

CONTENTS

11.1 FUELS

A substance that, when burned, produces heat and other forms of usable energy is referred to as fuel. Chemical and thermal energy are the most prevalent types of energy that are discharged into the surrounding environment. As a result of advancements in nuclear technology, it is currently possible to generate nuclear energy through either the fission process or the fusion process. Both of these processes are known as nuclear reactions. Burning fuels results in the production of thermal energy, which may be put to use in a variety of contexts, including the preparation of food, the heating of buildings, and a variety of commercial and industrial processes (Vasapollo et al., 2011).

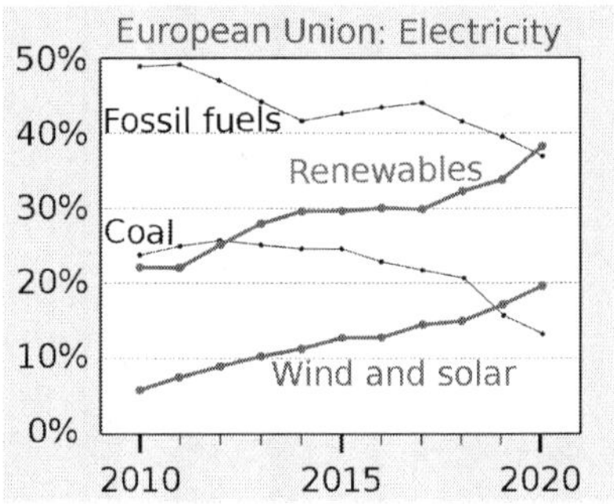

Figure 11.1: Fuels.

Source: By RCraig09 - Own work, CC BY-SA 4.0, https://commons.wikimedia.org/w/index.php?curid=99264795

A motor is utilized to convert the thermal energy into a form that can be used mechanically. This process is referred to as "thermal conversion." The same process occurs when we put fuel in our vehicles to get them to work. For there to be an optimal performance from engines, a certain kind of oil known as fuel oil is required.

Then there is what is known as the gasoline that our vehicles run on. Fuel is essential to our survival. Every cell in the body needs a supply of energy to function properly. The organic substances they consume, such as carbs, lipids, and proteins, supply them with the energy they need to function. Cellular respiration is the process that takes place whenever fuel is used up within the cell. Because it supplies the body with the many chemical elements that are necessary for it to carry out its functions in an efficient manner, food is sometimes referred to as the "fuel for the body." Fuels are concentrated sources of energy that are used to provide a variety of energy services, including heating, transportation, and the generation of electricity. Fuels may be broken down into hydrocarbons and organic compounds. Even though the majority of fuels get their energy from the sun in the end, many people still consider fuels to be primary sources of energy. When people discuss energy conservation, they almost always refer to reducing the amount of energy that they use.

The generation of about 95 percent of the world's total core energy (as opposed to primary energy flows) is attributable to primary fuels. Primary energy flows are a different concept. These fuels are put to use in power plants, engines, heaters, and any other machinery that is capable of producing either work or heat. In addition to this, they have a variety of additional applications to which they are put to use (Turney et al., 2012). When looking at the generation of electricity, the percentage is very different; fuels supply 85 percent of the primary energy that is required for the production of electricity.

11.2 TYPES OF FUEL

Because of this, the quantity of energy that can be extracted from burning fuel is greatly reduced. Even though the great majority of fuels do not come from renewable sources, it is feasible to identify them in sufficient numbers for them to be called sustainable. Consider the wind as an example of a primary energy source apart from fossil fuels.

Nuclear fuels, biofuels, and fossil fuels make up the majority of primary fuels. It is common practice to change primary fuels into molecules that are chemically distinct from those that were originally acquired from a natural source. To put it another way, crude oil is a primary source of fuel and is fractionally distilled to produce more consumer-friendly products.

Although kerosene and diesel are both alternative fuels, they are distinct from gasoline in that they are produced from primary sources of energy (Aramwit et al., 2010). These fuels are used in addition to the main fuels. These secondary fuels are the outcome of the processing of a natural resource, and one may consider them to be a sort of energy money. Crude oil is routinely converted into secondary fuels, which are more readily combustible in engines. As a result, crude oil may be used to its full potential in terms of energy production.

Fractional distillation, on the other hand, requires blending multiple fuels, such as methane, butane, and propane, and then separating them back into their original states. As a result, the world's most valuable energy resource is created. It's considered an energy currency since hydrogen isn't found in great amounts naturally on Earth, but it can be produced chemically from water or methane (among other things). It's possible to employ other sources of energy as well.

Fuels can differ substantially in terms of their energy density, cost, and environmental effect, making it difficult to compare them. Although uranium has a far higher energy density than fossil fuels, it is much more expensive to mine and produces. Because of the wide range of possible applications, it is impossible to directly compare the energy density and environmental effect of various fuels.

11.3 SOLID FUEL

Energy is generated by the burning of these solids. Coal, charcoal, soot, and wood are just a few types of solid fuels. Without a doubt, they were among the earliest sources of human fuel. They acted as a source of energy for the emergence of fire throughout time. They're still a crucial part of many families and companies today, throughout the world (Aidas et al., 2014). Charcoal is still a significant fuel source for all sorts of industrial machines as well as the power industry when it comes to cooking food and keeping warm in the home.

11.4 LIQUID FUEL

These are the sources of mechanical and kinetic energy that our civilization relies on. For most liquid fuels, such as crude oil, high temperatures and pressures are used to break down fossilized plant and animal remnants over a lengthy period. Ethanol and hydrogen fuel, both of which are liquid biofuels, can also be used. The conveyance and consumption of these fuels are frequently simple.

11.5 FUEL GAS

Fuel gas, as its name implies, is a gaseous fuel that is commonly found in nature. Several gases, such as methane, carbon monoxide, and propane, fall under this category. When it comes to transporting them, they have a significant benefit. The downside is that they can leak from the pipes, which is something to avoid at all costs. So, you may use your kitchen's existing gas lines to bring CNG gas inside your home. This sort of gasoline can also be referred to as "domestic fuel.

11.6 BIOFUEL

Solid, liquid, and gaseous forms of biofuel are all possible. A biomass-based product is the only requirement. It is a renewable fuel that is generated by living beings in its most basic form. For example, sugarcane residue may be used to make ethanol.

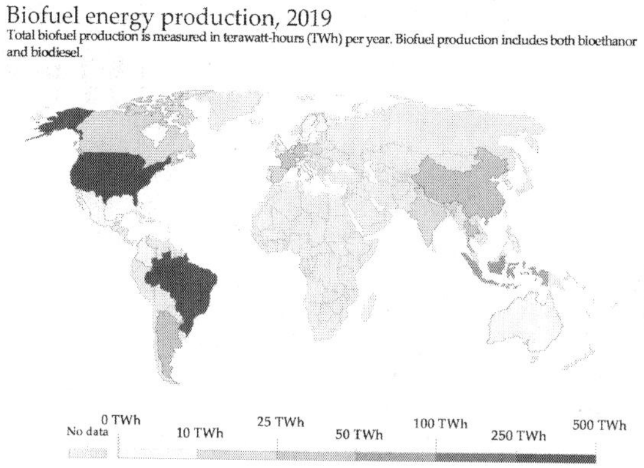

Figure 11.2: Biofuel.

Source: By Our World In Data - https://ourworldindata.org/grapher/biofuel-production, CC BY-SA 4.0, https://commons.wikimedia.org/w/index.php?curid=97378658

11.7 FOSSIL FUEL

These are the most common kind of energy sources. In addition, it is impossible to duplicate them. Fossils were generated millions of years ago when petrified animal and plant parts were exposed to high pressure and heat in the planet's core. Because we rely so heavily on these resources, this is a major problem for mankind. It is important to note that many distinct forms of fossil fuels may be found in the world today. Natural processes acting on the organic molecules left behind after photosynthesis, which began between 4.0 and 2.5 billion years ago, created all of them (4.0 billion to 2.5 billion years ago). Carbonaceous material was mostly generated by algae and bacteria during the Pre-Devonian Period (419.2 million to 358.9 million years ago).

Plants, on the other hand, provided the vast bulk of the carbonaceous material that existed at that time and in the years that followed. Any fossil fuel may be burnt to generate heat, either in the open-air or with the assistance of atmospheric oxygen_(Thomford et al., 2018). Direct use of the heat, such as in furnaces found in houses, or the production of steam, which may subsequently be used to power electrical generators, are both viable options. If fossil fuel is burned in a gas turbine engine of a jet plane, that heat is used to enhance the pressure and temperature of combustion products, which in turn generates enough power for the motor to operate. A steady increase in fossil fuel consumption has been occurring since the Industrial Revolution began in Great Britain at the end of the 18th century. This is a trend that is likely to continue. They presently provide more than 80% of the energy used by the world's wealthiest countries. Even though new sources of fossil fuel are continuously being discovered, the world's principal fossil fuel resources are in short supply.

Several variables make it impossible to estimate the entire amount of fossil fuels that may be recovered at a profit, including the pace of use, future value, and the development of new and improved technology. Fossil fuel resources smaller and more difficult to access may now be extracted at acceptable costs because of advances in technology like hydraulic fracturing (commonly known as "fracking") and rotary drilling, as well as directional drilling. In turn, this has increased the total amount of salvageable material. Certain oil corporations began producing heavy oil as well as liquid petroleum derived from tar sands and oil shale when traditional (light to medium) oil stocks depleted. This was necessary due to the depletion of conventional oil supplies. One of the most obvious consequences of burning fossil fuels is the emission of carbon dioxide (CO2). The increased use of fossil fuels in industry, transportation, and building development has resulted in a massive emission of carbon dioxide into the atmosphere.

11.8 ORGANIC CHEMICALS

In the first stages of the heavy chemical industry, inorganic chemistry was utilized as a method for the investigation of compounds that did not include carbon or any of its constituents. As was said before, inorganic chemistry does include the presence of carbonates. In the light chemical sector, organic chemistry, which concentrates on carbon molecules and includes hydrocarbons as examples, is employed in a manner that is not distinct from how it is used in other industries (hydrogen-carbon pairs). During a brief period in the 1960s, several different chemicals, such as vinyl chloride and benzene, were categorized as "heavy organic compounds." In terms of their chemical composition, benzene, phenol, toluene, and the xylenes are all classified as belonging to the same category of aromatic organic compounds (Arif et al., 2010).

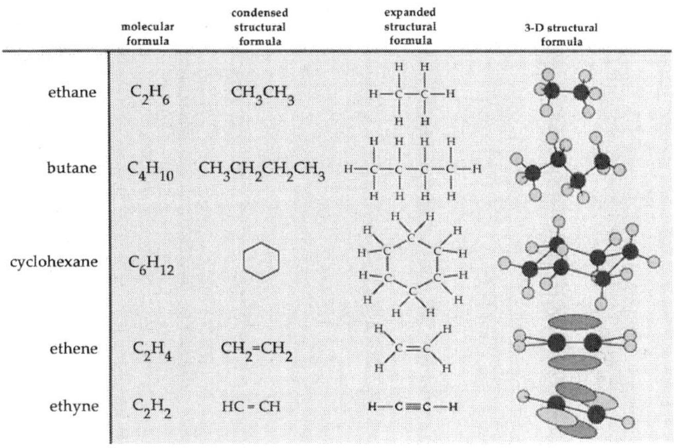

Figure 11.3: Organic Chemicals.

Source: https://cdn.britannica.com/99/7399-004-3A360F3D/structures-compounds-formulas.jpg

To a large extent, the chemical properties of aromatic compounds are comparable to those of benzene.

Compounds made of organic matter can be broken down into the following groups of elements:

11.9 BENZENE

The structure of the aromatic hydrocarbon benzene mimics that of a closed ring and consists of six distinct sides. An atom of hydrogen is positioned inside of a carbon atom at each of the hexagonal corners. Every molecule of benzene has twelve hydrogens (H) atoms and a total of one hundred and sixty-six carbon (C) atoms, making the total number of atoms one hundred and sixty-six. Since the beginning of civilization, benzene has been an important component of society in many different ways. The carbonization of coal results in the production of coke, combustible gas, and a wide variety of other byproducts, including benzene. Carbonization of coal was used as a method of creating lighting gas in England around the beginning of the nineteenth century. This process was introduced at that time. Although it is still utilized in some regions, the usage of natural gas is becoming more and more widespread. The manufacturing of iron and, by extension, steel, requires the use of metallurgical coke, which is created through the process of carbonization. These two metals derive from iron, which is their parent element (with minor alterations). However, the amount of benzene that is produced as a result of the carbonization process is not enough to satisfy the requirements of the market. The carbonization of one ton of coal results in the production of only two to three pounds of benzene (0.9 to 1.35 kg).

During World War I, there was a major scarcity of aromatics as a result of the high demand for trinitrotoluene (TNT), which led to a severe lack of aromatics. This shortage was caused by the high demand for TNT. TNT was by far the most commonly used explosive in those days. The extraction of toluene from crude oil calls for the utilization of a variety of processes. After World War II, there was a greater demand for benzene and other aromatics than the existing supply of metallurgical coke could meet (Barbatti et al., 2014). This caused a shortage in the market. At this point, the vast majority of aromatics are derived from petroleum.

11.10 TOLUENE

Toluene is distinct from benzene in that a methyl group has been situated in the place of one of its hydrogen atoms in its chemical structure (CH_3). Toluene is distinguishable from benzene due to the peculiar carbon-hydrogen bonding that it possesses. Because xylenes have two methyl groups attached to the benzene ring in a variety of positions, aromatics can largely be interchanged with one another. Actuality, benzene is produced as a byproduct of the chemical reaction that occurs when the methyl group of toluene is removed.

Because of their inherent antiknock properties, each of these hydrocarbons is an excellent candidate for use as an additive in gasoline. In addition to that, solvents like toluene and others are utilized. The phrase "in the capacity of a solvent" is frequently used in the chemical industry, and its meaning can refer to a great variety of different activities. Many different processes involve dissolving organic molecules, some of which include the production of coatings, adhesives, textiles, medicines, inks, and photographic film. One service that is provided to customers directly and immediately is dry cleaning (although the solvent used here is not toluene, but other hydrocarbons or chlorohydrocarbons). Toluene is a chemical that can be discovered in a wide variety of products (Bacellar et al., 2015). Polyurethane plastics and elastomers have the potential to contain this substance.

11.11 XYLENE

The three isomeric forms of xylene each contain a compound that differs from the benzene molecule in that one of its hydrogen atoms has been replaced by a C_2H_5 atom. Given how challenging it is to distinguish between isomers, it is surprising that so many different techniques for separating them have been developed. The prefixes ortho-, meta-, and para- are added to the word "xylene" to differentiate between three distinct isomers of benzene that are based on how the two methyl groups displace the hydrogen atoms. These prefixes are used in place of the term "xylene." A primary component of ortho-xylene is required for the production of phthalic anhydride, which is an essential step in the manufacture of a wide variety of coatings and polymers. Although meta-xylene is the isomer with the lowest potential return on investment, it is used in the manufacturing of paints as well as polymers. Polyesters can be derived from para-xylene, and once produced, polyester fibers can be marketed and sold under a wide variety of registered brand names.

Methane (CH_4), ethane (C_2H_6 or H_3CCH_3), and propane are three of the most fundamental chemical compounds. Methane has the formula CH_4 ($H_3CCH_2CH_3$). It is necessary to subject them to heat treatment to "crack" them and release the unsaturated hydrocarbons, which can then be used to produce new compounds. They have one or more double or triple valence bonds connecting the carbon atoms, and they have a lower hydrogen content than saturated hydrocarbons. In industry, some of the unsaturated hydrocarbons that are used most frequently include butadiene, acetylene, ethylene, and propylene. Other examples include propylene and ethylene.

11.12 ETHYLENE

Ethylene, which is among the most widespread chemical compounds, can give rise to the formation of oxygen. Many other products are designed using ethylene.,. By placing some of the most important chemicals in a box, it is possible to gain a better understanding of the intricate bonds that exist between them This gives you the ability to see how they are all connected (Bennett et al. 2010). These bonds would become clear after conducting a thorough investigation of this box. It is possible to produce acetylene from calcium carbide, although this does not mean that acetylene is the only source from which these molecules can be derived.

11.13 ACETYLENE

Lime, coke, and electricity are the primary ingredients that go into the manufacturing of calcium carbide. In a country that has an abundance of hydroelectric power but no petroleum reserves, calcium carbide is the most common and reliable source of acetylene. The production of acetylene is dominated by several countries, including Japan, Poland, the Soviet Union, and several others. The reaction of calcium carbide with water results in the production of acetylene. This method, which results in a very intense flame, can be utilized to accomplish the production of acetylene on a more modest scale. This can occur due to the characteristics of the method. Figure 3 illustrates the large-scale production of acetylene, which is done to undergo a chemical transformation. Oxyacetylene welding is utilized because of the significant increase in temperature that occurs when acetylene is combined with oxygen in the welding process. There is fierce competition between acetylene and ethylene for use in various chemical industrial applications.. In the 1950s, acetylene was a popular option when it came to serving as a raw material for the synthesis of various chemicals. Because it is now possible to purchase ethylene at lower prices, the demand for acetylene as a raw ingredient has begun to decrease. Calcium carbide is the starting material for the production of acetylene, but this chemical has a variety of other uses as well. Calcium cyanamide is a fertilizer and herbicide that can also be used to make melamine, which is a raw material in the creation of several modern polymers (Congrains et al., 2013). Calcium cyanamide is produced when nitrogen combines with calcium cyanamide to produce calcium cyanamide. Calcium cyanamide, in addition to being used in the manufacture of melamine, may also be put to use in the production of advanced.

11.14 PROPYLENE

Propylene is produced at a volume that is significantly lower than that of ethylene, and its primary application is in the chemical industry. Ethylene is the chemical that is most commonly used. It is frequently found as an ingredient in cleaning products. It does this by producing compounds, which are then incorporated into the gasoline as additives, to prevent the engine from knocking. It is also possible to polymerize it into a substance that is analogous to polyethene and can be utilized for purposes that are comparable to those of polyethene. Due to the high level of effectiveness, it possess, polypropylene makes an excellent material for carpeting.

11.15 ELASTOMERS

This substance is used in the production of various types of plastics, including polymers and elastomers, both of which use butadiene as a chemical building block). In the beginning, people thought that elastomers were just synthetic rubbers, but now we know that this is not the case. It was immediately understood that what had been generated as a supplement rather than a replacement. Synthetic substitutes frequently come in several forms, and it was immediately understood that what had been generated as a supplement. Some of these types of rubber were superior to natural rubber in terms of performance, while others were superior in other respects.

Because of the disruption in the supply of tropical rubber, Germany became interested in developing a synthetic compound that could be used in the manufacturing of automobile tires during World War I. This was because the supply of tropical rubber was being disrupted. Following the end of the war, people's interest in the topic did not fade away. To avoid the development of flat spots on the tires, the vehicle needed to be jacked up every time it was parked for an extended period. As a direct consequence of this, a synthetic rubber that is appropriate for use in tire tires has been developed. In the years leading up to World War II, both Germany and the United States were active participants in the research and development of a wide range of elastomers (Campos & Vasconcelos, 2010). This particular copolymer, which was composed of 75% butadiene and 25% styrene, was the one that turned out to be the most significant. Tires showed a much more positive response to this particular copolymer than they did to the others. This synthetic was first known by the name Government Rubber Styrene (SBR), but today it is more commonly referred to by its abbreviated form, GR-S (Government Rubber–Styrene) (styrene-butadiene rubber). It is without a doubt the synthetic substance that is manufactured the most frequently. In some respects, it is superior to natural rubber, whereas, in others, it is not as good as natural rubber. It is commonly used in conjunction with a variety of other types of rubber.

11.16 POLYMERS

Any substance or material that is made up of many smaller molecules or macromolecules that repeat itself is referred to as a polymer. Polymers, both man-made and naturally occurring, play a vital role in our daily lives. This is because of the wide range of characteristics that each kind has. From man-made plastics like polystyrene to biopolymers like DNA and proteins, polymers play an important role in the structure and function of living things. Polymers include synthetic plastics like polystyrene (Colussi et al., 2013). Natural and man-made polymers can be made by combining a large number of smaller molecules, known as monomers, into one larger molecule. Physical properties such as toughness and high elasticity, as well as the tendency to form amorphous or semi-crystalline structures rather than crystals can be found in large molecule compounds, which have large molecular masses. Because of these features, large molecule compounds are distinguished from small molecule ones.

11.16.1 Natural Polymers: Organic and Inorganic

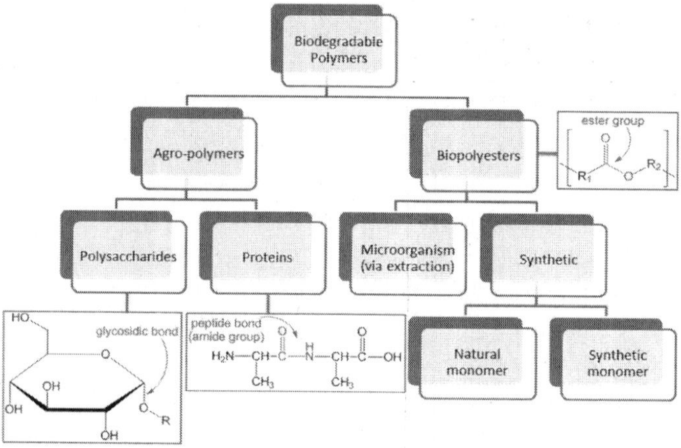

Figure 11.4: Natural polymers: biodegradable polymers organization.

Source: By 5402013SD - Own work, CC BY-SA 3.0, https://commons.wikimedia.org/w/index.php?curid=31592630

In living things, organic polymers play an important role in providing structural elements and facilitating critical life processes. Polymers make up the solid parts of all plants, to cite but one example. Cellulose, lignin, and other resins all fall under this umbrella. Because it is composed of sugar molecules, cellulose is known as a polysaccharide. Lignin is made up of a highly intricate polymer network that spans all three dimensions. The production of wood resins often involves the usage of hydrocarbon isoprene. Rubber is an additional well-known example of an isoprene polymer.

Proteins, which are polymers of amino acids, and nucleic acids, which are polymers of nucleotides, are two additional major natural polymers. Nucleic acids are complex substances that consist of nitrogen-containing bases, sugars, and phosphoric acid, while proteins are polymers of amino acids. Nucleic acids are responsible for disseminating genetic data throughout the cell. Starches are naturally occurring polymers that are predominately made up of glucose and serve as vital sources of nutritive energy for plants (Chen et al., 2015). The minerals diamond and graphite are both examples of inorganic polymers that occur naturally. Carbon is present in both of them. Because of the three-dimensional network of carbon atoms that gives a diamond its structure, a diamond has a very high level of hardness. Carbon atoms are joined together in planes in graphite, which is used as a lubricant and in the "leads" of pencils. Graphite has carbon atoms that can glide over one another.

and 15,000 are used in the production of films, molded items, and fibers such as Dacron.

Casein, which can be found in milk, and zein, which can be found in corn (maize), are two examples of the naturally occurring proteins that make up polyamides. Polyamides are used in the production of plastics, textiles, adhesives, and coatings. The thermosetting urea-formaldehyde resins are a type of synthetic polyamide that can be found in some products. Coatings, adhesives for textiles and paper, and molded goods production are some of the industries that make use of them. In addition to this, nylon resins, which are a type of polyamide resin, are an absolute necessity. They are not flammable, do not emit toxic fumes, can withstand high temperatures and abrasion, and can even be colored. In addition to the well-known application of using them as fibers for textiles, they also have several other potential applications.

An additional important category of synthetic organic polymers is represented by linear repetitions of the urethane group. Polyurethanes are used in the manufacturing of a variety of products, including coating bases, soft and stiff foams, and elastic fabric spandex.

CHAPTER 12

CHEMICAL EQUILIBRIUM

CONTENTS

12.1 INTRODUCTION

In a chemical process that can be reversed, equilibrium is said to have been attained when there has been no change in the overall quantity of either the reactants or the products since the beginning of the process. When the products of a reaction can swiftly interact with the reactants that were utilized in the process, we refer to that type of reaction as being reversible. This guarantees that the reaction can be carried out in the opposite direction from what was originally intended. Since there is neither an increase nor a decrease in the total amount of chemicals that are involved when two chemical processes operate at the same rate or velocity, there is no change in the total amount of chemicals that are involved (Chaplin, 2006). At this stage, the reaction is complete since all of the reactants have been transformed into products to the greatest extent that is feasible under the conditions that are now present in the reaction. This means that the process has reached its conclusion.

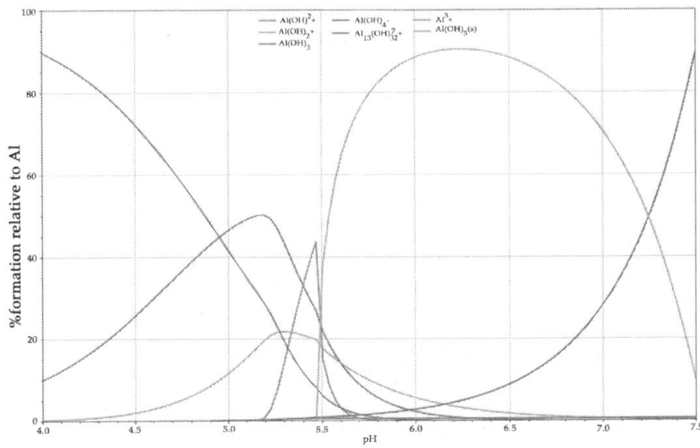

Figure 12.1: Chemical Equilibrium.

Source: By Krishnavedala - Own work, CC0, https://commons.wikimedia.org/w/index.php?curid=34342616

Some chemical reactions don't ever seem to reach their full potential, even if they can. The great majority of chemical reactions are reversible, which means that the sequence in which the reactants and products are produced may be reversed whenever the experimenter so chooses. An equation representing a reversible reaction will have two arrows, one of which will go forward while the other will point backwards. This is because a reversible reaction contains both forward and reverse processes. At the most fundamental level, the two processes that are opposed to one another are competing against one another to achieve their respective goals. Chemical equilibrium is a type of dynamic equilibrium that allows for the potential of chemical reactions to take place inside its limits. This makes a chemical equilibrium different from other types of equilibrium.

There are many distinct chemical processes, and each one may be manipulated to go in either the forward or the backward direction, depending on the context in which it is being carried out. The state of equilibrium that occurs between two processes that are hostile to one another and that take place during a reaction that may be reversed is referred to as "chemical equilibrium," and the phrase "chemical equilibrium" refers to that condition. Both forward and reversal reactions, represented by the letter D, take place concurrently throughout this dynamic process. When a chemical process has reached equilibrium, the ratio of the concentrations of the reactants to the concentrations of the products continues to be maintained at the same value. As soon as there is equilibrium, there will be reactions in both the forward and the backward directions (Chemat et al., 2012). Once equilibrium has been reached, the principles of kinetics limit the rate at which it may be re-established once it has been lost. If a container did not have any vents or openings, the following series of events would happen to it: 2HI is the same as $2H_2$ plus I_2, alternatively, it may be written as $H_2 + I_2 \rightarrow 2HI$

Hydrogen iodide is a product that is formed through a chemical process that involves the combination of hydrogen and iodine. This product's name comes from the fact that it is made through this procedure. Since the equation tells us that all of the H_2 and I_2 have joined to generate HI, we are compelled to conclude that the reaction is complete given that it specifies that this has occurred.

On the other hand, this is not the case in any way, shape, or form in the least! In addition to this, a chemical reaction that operates in the opposite direction takes place: $2HI \rightarrow H_2 + I_2$

The interaction between H_2 and I_2 results in the formation of 2HI as the final product.

It has the effect of reversing the outcomes that were caused by the initial reaction. When it comes to velocity, the reaction in backward motion has now caught up to the reaction in forwarding motion, which means that the opposite direction is moving faster than it was before. When this occurs, the reaction has reached a state known as chemical equilibrium, also known as equilibrium (also known as equilibrium; plural equilibria), which is the point at which forward and backward processes balance each other's progress (Sneskov & Christiansen, 2012). Equilibrium is the state in which a reaction has achieved a state known as chemical equilibrium, also known as equilibrium (also known as equilibrium; plural equilibria). At this moment, currently indicates that any global reaction that has been going on constantly for a long time has come to an end.

It is usual practice to depict equilibrium with a double arrow when two processes cannot coexist with one another. This is because the processes cannot work together. The following outlines each step in these processes:

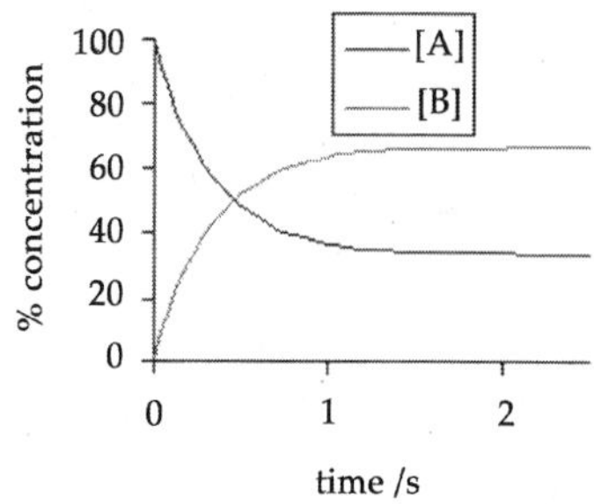

Figure 12.2: Chemical Equilibrium.

Source: By Petergans at English Wikipedia - Own work (Original text: I (Petergans (talk)) created this work entirely by myself.), Public Domain, https://commons.wikimedia.org/w/index.php?curid=23023862

In such a case, we would be dealing with H2+ I2⇄2HI

The answer can travel in either way, as seen by the figure›s two arrows, which point in opposing directions. This can be noticed by looking at the diagram. It is of the highest significance to keep in mind that the reaction as a whole has to retain itself in a state of equilibrium throughout its whole.

When contemplating equilibrium, one thing to bear in mind is that reactions continue to occur in both the forward and backward directions. This is an important aspect of equilibrium. You must keep this in mind. Because the two reactions are occuring at the same moment, they can nullify one another, which results in there being no change overall. It is more accurate to refer to the condition of events as dynamic rather than as a static one when describing the circumstance of chemical equilibrium (Stuppia et al., 2012). Because both processes are occurring at the same time, it is possible to define a state of equilibrium in the other direction. This is possible because both processes are occurring simultaneously. Take, for example, the situation in which you wish to convey the concept that equilibrium is a condition of balance. Consider the scenario.

In such a case, we would be dealing with the $H_2 + I_2 \rightleftarrows 2HI$

In the same manner, as if a different term had been chosen to express the same equilibrium position.

$H_2 + I_2 \rightleftarrows 2HI$

Having said that, you will not be able to complete this step until the reaction has reached equilibrium.

It ought to be rather evident to us that once equilibrium is reached, the reaction will terminate. Once equilibrium is attained, the reaction will come to an end. In chemistry, the term "equilibrium" refers to a condition that is not static but rather one that is always changing. Even when a situation of equilibrium has been reached, both backward and forward reactions continue to take place. On the other hand, when a reaction has reached equilibrium, the relative concentrations of the reactants and the products do not change because the reaction rates do not change (Rhoads et al., 2012). This is because the concentrations of the reactants and the products are held constant.

One is obligated to ensure that the system is completely sealed, which implies that no substances may enter or leave the system. This is a responsibility that one must fulfill. An equilibrium that is stable does not exist in any universe. Reactions are taking place in both the forward and the backward directions, even if we are oblivious of what is occuring. The forward and backward reactions must have a frequency of occurrence that is comparable to one another.

It is not necessary for there to be an exact correlation between the numbers of products and reactants in a chemical reaction. After the condition of equilibrium has been attained, there will be no change in the quantity of the reactants or the products. When referring to this concept, the word "equilibrium" refers to the condition of equilibrium that is maintained during a chemical reaction.

Phase equilibrium and solution equilibrium are two more types of equilibrium that can exist, to mention a few instances of each. It is said that equilibrium has been reached in a system when a material is in a state that is midway between two other states. If we take the example of a flask that is filled with water and has a stopper on it, we can determine whether or not the flask is in equilibrium by determining whether or not the rate of evaporation corresponds with the rate of condensation (Rastogi & Sani, 2011). When a solid substance is completely submerged in a solution that is at its maximum concentration, it is said to have reached an equilibrium in the solution. At this current time, the rate of recrystallization and the rate of dissolution are both equal. The principles of equilibrium can be applied to any of the processes that can be reversed, even though each of these processes involves a unique set of transitions of some kind.

The most vital function that oxygen-carrying red blood cells are responsible for performing in the body is providing oxygen supply to various tissues. Cells can't carry out the metabolic processes that they are responsible for when the oxygen supply to the cell is inadequate. A molecule known as hemoglobin is responsible for delivering oxygen to the cells of the body. Hemoglobin is found inside red blood cells. Carbon monoxide binds to hemoglobin with a significantly higher level of tenacity than oxygen does, and as a result, toxic levels of this gas reduce the amount of oxygen that can enter the cells. The patient's body will be able to eliminate the carbon monoxide that has accumulated if they continue to take in pure oxygen. The reaction at equilibrium shifts to the right when an excessive amount of oxygen is introduced into the system:

$$Hb(CO)_4(aq)+4O_2(g)\rightleftharpoons Hb(O_2)_4(aq)+4CO(g)$$

12.2 EQUILIBRIUM CONSTANT

Imagine for a moment that the reactants aA and bB are involved in an irreversible reaction that results in the production of the products cC and dD. In this equilibrium, the coefficients of each component of the formula are indicated using lowercase letters. The formula itself is presented above.

$$aA+bB\rightleftharpoons cC+dD$$

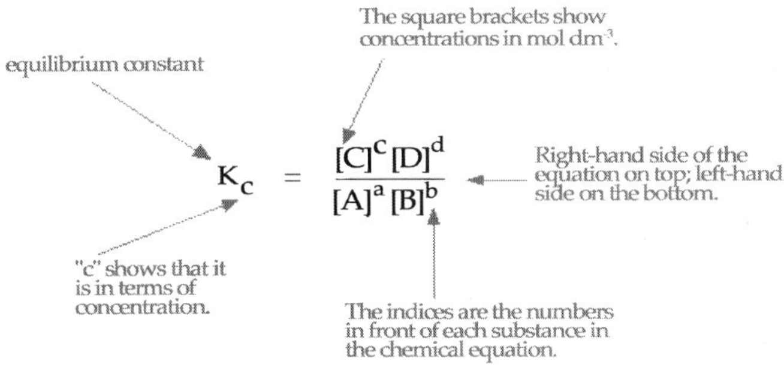

Figure 12.3: Equilibrium Constant.

Source: https://chem.libretexts.org/Bookshelves/Physical_and_Theoretical_Chemistry_Textbook_Maps/Supplemental_Modules_%28Physical_and_Theoretical_Chemistry%29/Equilibria/Chemical_Equilibria/Kc

Because the rates of both forward and backward reactions are identical when there is equilibrium, the concentrations of all molecules stay constant. This is something that we have demonstrated. As a consequence of this fact, the concentration ratio of any reaction that has reached equilibrium will continue to be the same over time. The word "equilibrium constant," abbreviated as "Keq," is a phrase that is used to define a relationship between the concentrations of the product and the concentrations of the reaction reactants in a process (Dubernet et al., 2010). In the chemical equation, the coefficients that correspond to the various balanced concentrations are multiplied.

Experimentation is the only way to learn the value of the equilibrium constant for a reaction. Since the value of the equilibrium constant does not vary, it was established in the preceding paragraph that the equilibrium site for a given reaction is unaffected by the concentrations that were present at the beginning of the reaction. This was shown by the fact that the value of the equilibrium constant did not change. The temperature is a vital aspect to take into account in this case. The condition is not stable as equilibrium is defined as having equal forward and reverse reaction rates; as a result of this, equilibrium cannot exist in the current state. Because of the variability in reaction rates, the equilibrium constant will fluctuate in reaction to fluctuations in temperature. When a Keq is utilized as a reactant in a reaction, the temperature at which it is carried out is very vital to define.

When Keq is larger than 1, the numerator is more than the denominator, which implies that the concentration of the product is greater than the concentration of the reactant. The fact that Keq is smaller than one implies that the denominator comprises a higher number of reactants than does the numerator (products). When Keq = 1, the concentrations of the reactants and the products are practically similar to one another.

A system is regarded to be in equilibrium when the concentrations of the reactants and products do not vary over time. The nature of this situation may be defined using a broad number of observable factors, including the stability of the pressure and density. The balance that is generated as a result of physical processes is referred to as the situation of physical equilibrium. A chemical reaction is said to be in a state of chemical equilibrium when its components are in a condition of balance with one another.

The system's temperature, pressure, and concentration can all affect how the equilibrium is maintained. For instance, if even a single one of these components is altered, the equilibrium of the system will be thrown off, and the system will need to adjust to reach its former condition. In this and the next sections, we will cover some of the most essential elements that influence equilibrium.

12.3 THE CHEMISTRY BEHIND REACHING AN EQUILIBRIUM

A system is regarded to be in chemical equilibrium when the concentrations of the reactant and product do not vary over time and the characteristics of the system do not change in any manner. Another evidence of this condition is that there is no obvious change in the system's properties. The system is deemed to be in a condition of equilibrium so long as the forward and backward reactions continue to move at the same speed. The state of equilibrium is obtained when the rate of the reaction moving forward is similar to the rate of the reaction going backwards. As a consequence of this, the term is typically used to define a system in which the concentrations of both the reactant and the product are maintained throughout the experiment (Dufrêne et al., 2017). When there is no longer any change in the system's properties, the system will have achieved its fixed state. It is stated that the concentrations of the reactants and products have reached a condition of dynamic equilibrium when the forward and reverse processes utilize rates that are similar to one another.

12.4 THE FACTORS THAT AFFECT EQUILIBRIUM

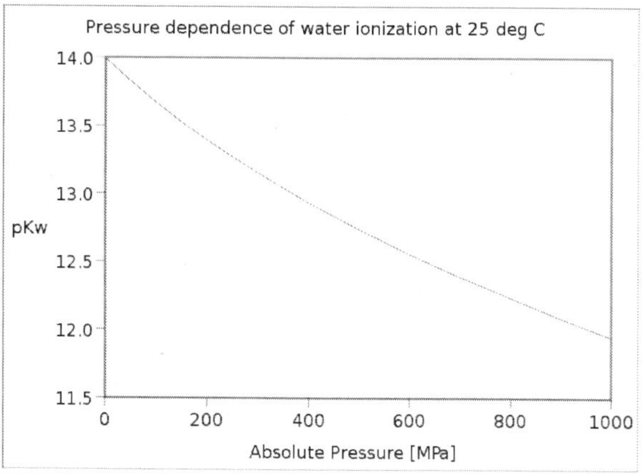

Figure 12.4: The Factors That Affect Equilibrium.

Source: By Stan J Klimas at English Wikipedia - Transferred from en.wikipedia to Commons by Stan J Klimas., Attribution, https://commons.wikimedia.org/w/index.php?curid=5683540

Le-Chatelier and Braun, two French scientists, came up with several general assertions about the influence that modifications in temperature, concentration, or pressure have on the equilibrium state of a system. For instance, if even any one of these components is altered, the equilibrium of the system will be thrown off, and the system will need to adjust to reach its former condition. This oversimplification is referred to as "the Le-Chatelier Principle," and its name originates from its namesake. The Le-principle asserts that when the circumstances of an equilibrium change in a system, the equilibrium will adjust in such a manner that the effect of the change will be diminished or counteracted. This is because the equilibrium will adjust in such a way as to maintain the system's overall stability.

When a system is in equilibrium, even a tiny adjustment in the concentration, pressure, or temperature can have a large influence on the state of the system. This concept acts as the controlling force behind a fundamental principle in both chemistry and physics (Ebbesen, 2016). The system's temperature, pressure, and concentration can all affect how the equilibrium is maintained. In the following sections, we will talk about some of the most fundamental factors that affect equilibria.

12.5 EFFECT OF A CATALYST

In the presence of the catalyst, there does not appear to be any appreciable disruption to the equilibrium. This occurs because the catalyst can favor both forward reactions and backward reactions. While the system is in a condition of equilibrium, both the forward and reverse rates, as well as the forward-to-reverse rate ratio, stay unchanged. This holds about the ratio of reactants to products as well. This demonstrates that the existence of a catalyst does not directly change the condition in which the equilibrium is maintained. [Cause and effect]

This suggests that the catalyst itself does not have any influence on the current state of the chemical equilibrium, which is the conclusion that can be reached from the information presented here. It simply helps to cut down on the amount of time that must pass before reactions can take place (Hasanuzzaman et al., 2013). Depending on the circumstances, it may be feasible to employ a catalyst in either the forward or the backward direction of a process to speed up the process. After it has gone on its voyage, the reaction will achieve equilibrium at a faster rate as a direct result of this, which is a consequence that cannot be avoided or avoided directly.

Both catalyzed processes and non-catalyzed processes result in the same total number of reactants and products when the system is in a state of equilibrium. Catalysts shorten the length of time needed for a reaction to attain its final aim by enabling it to bypass a stage in the process in which a higher quantity of energy is required. This allows the reaction to use less energy overall.

12.6 EFFECT OF ADDITION OF AN INERT GAS

The addition of a gas that is not active may have consequences that were not expected. The introduction of inert gas into the equilibrium system may have the following effects, although they are not guaranteed to occur under any given set of circumstances: The overall pressure of the system can be increased by including a predetermined quantity of a non-reactive gas in the mix. When a constant quantity of inert gas is added to an existing equilibrium, the equilibrium will be thrown off. On the other hand, during the process, there will be no change in the concentrations of either the reactants or the products.

As a consequence of this, the equilibrium will not be disturbed even if these requirements are not fulfilled. The volume of the system is increased by the addition of an inert gas while the pressure is maintained at the same level. This is the effect that takes place as a direct consequence of the inclusion of a pressured inert gas within the system. As a result, the likelihood that a certain volume containing the number of different reactants and products would comprise a significant number of moles is reduced. Because of this, the equilibrium will shift in a way that will make it easier to increase the total amount of moles of gas that are present in the system.

12.7 DYNAMIC EQUILIBRIUM

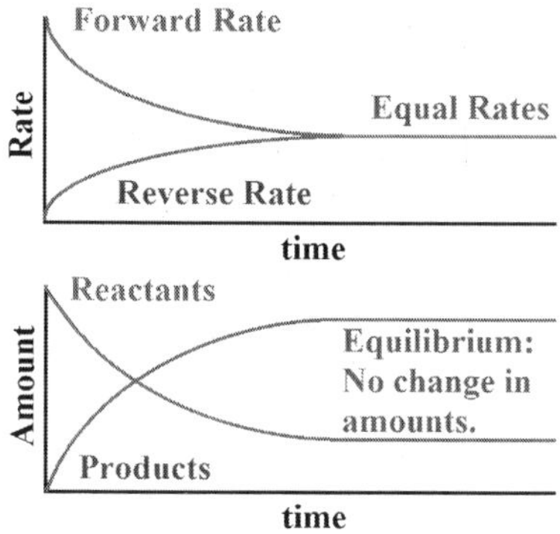

Figure 12.5: Dynamic equilibrium.

Source: https://encrypted-tbn0.gstatic.com/images?q=tbn:ANd9GcRoakkoJWQwSQ16sqgJTTn1c5lv6Euq5aMLuEtGT9n7MqDL9MI2OHATX737YKm02fuMGX8&usqp=CAU

Chemical equilibrium is not achieved in a static but rather in a dynamic manner. When a system is in equilibrium, both the forward and reverse reactions are occurring at the same rate in the same amount of time. As a result, the ratios of the reactants to the products remain constant throughout the reaction.

Here's an illustration of dynamic equilibrium in action.

For example, the production of ammonia from its two constituents, nitrogen and hydrogen, is an example of this. Specific experimental settings can be used to determine the equilibrium composition of NH_3, N_2, and H. Only nitrogen is capable of absorbing radioactive forms of nitrogen. In contrast to nitrogen, the radioactivity of ammonia increases as the process progresses (Huang & Zou, 2010). The rate of radioactivity increase for ammonia is equal to the rate of radioactivity decrease for nitrogen, as previously stated. A chemical reaction is in equilibrium when the rate of the forward reaction equals the rate of the backward reaction, according to experimentation.

12.8 EXAMPLES OF CHEMICAL EQUILIBRIUM

A container that has been hermetically sealed maintains a constant temperature for the vapors of hydrogen and iodine. Some of the hydrogen and iodine are converted into hydrogen iodide towards the end of the procedure. Because the reactant and the product have reached a state of equilibrium after reaching this dynamic point. For a long amount of time, hydrogen iodide will break down into molecules of hydrogen and iodine when held in an airtight container at the same temperature. Hydrogen iodide residue remains unchanged as a result of this. Contrarily, no difference is seen in the concentrations of hydrogen, iodine, or iodide of hydrogen in either experiment (Raies&Bajic, 2016). When a stage of dynamic activity has been reached, it is claimed that equilibrium has been achieved.

12.9 EXAMPLE OF A REVERSIBLE REACTION

If the same circumstances are applied to the vessel in both experiments, the equilibrium composition will be the same in both instances. Because of this, reactions can be expected from both sides. This event is known as a reversible reaction, and the symbol for it is.

12.10 FORMATION OF AMMONIA

When using a catalyst, ammonia production from nitrogen and hydrogen at 450 degrees Celsius and 250 atmospheres can be reversed at the same temperature and pressure conditions. The ammonia produced by the reactant molecules was responsible for 13% of the total output. A total of 87% of the ammonia will be converted to nitrogen in the same conditions, while the remaining 2% will be converted to hydrogen. As a result, it is an equilibrium reaction that can be reversed.

12.11 HYDROLYSIS OF BISMUTH CHLORIDE

Milky in appearance, bismuth oxychloride is the substance that emerges from the reaction that takes place when bismuth chloride is combined with water.

$BiCl_3 + H_2O \rightleftharpoons BiOCl + 2HCl$

When a hydrochloric acid solution is added to this milky solution, the milkiness is lost. Changes in environmental conditions can cause changes in the equilibrium, as shown in this example.

12.12 REVERSIBLE AND IRREVERSIBLE PROCESSES

When equilibrium is reached in a reversible chemical reaction, the specific heat does not vary. On the other hand, all irreversible processes, such as spontaneous chemical reactions, raise the value of total entropy over zero. Unchangeable equilibrium is reached when the overall heat change increases, such as when the iron is reduced in a stream or syrup is diffused by water (Pritchard et al., 2019). Thus, the entropy has increased to its maximum feasible value as a result of this equilibrium. There are no chemical processes or changes in entropy when a system is in equilibrium.

CHAPTER 13

ACIDS AND BASES

CONTENTS

Acids and bases, two chemical varieties, enjoy exchanging particles. Acids are chemicals that, when dissolved in water, produce hydrogen ions. Simply, every substance that accepts protons responds quickly to them. The providers of protons are referred to as acids. A base is a substance that contains both oxygen and hydrogen atoms. Two negatively charged molecules make form a hydroxyl group. Because of their quick reactivity to positively charged atoms, bases are also referred to as proton acceptors.

Because of their high reactivity, acids and bases are vital in chemical processes. Many other species rely on them as well, making them critical to our existence. Acids have a tart flavor, whereas bases have a bitter flavor (Pinzi& Rastelli, 2019). Because our tongues recognize these chemicals, we can taste the acidity in lemons and the bitterness in the chocolate. Although some of these flavors may be attractive, this sense is essential in recognizing potentially harmful chemicals.

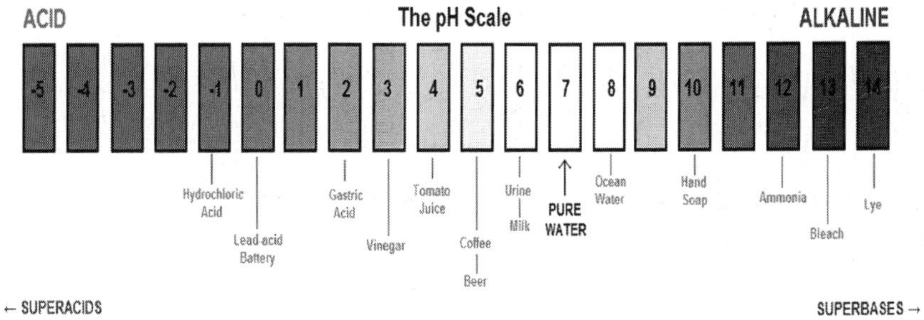

Figure 13.1: The pH scale.

Source: https://en.m.wikipedia.org/wiki/File:PH_scale_2.png

Acids and bases have a far larger role in the seas. Marine mollusks use chemical components found in water to construct their shells. To work correctly, shark nostrils require a specific pH level in the water. Carbon dioxide is released into the atmosphere as a result of the combustion of fossil fuels, causing the seas to become more acidic. Animals have a more difficult time forming shells in acidic water. The pH scale may be used to assess the acidity or basicity of a material. This scale ranges from 0 to 14. The pH of pure water is seven (Kalra et al., 2016). The pH of all acidic substances, including lemon juice and battery acid, is less than seven. The pH of bases such as bleach, oven cleaning, and even your blood is more than seven.

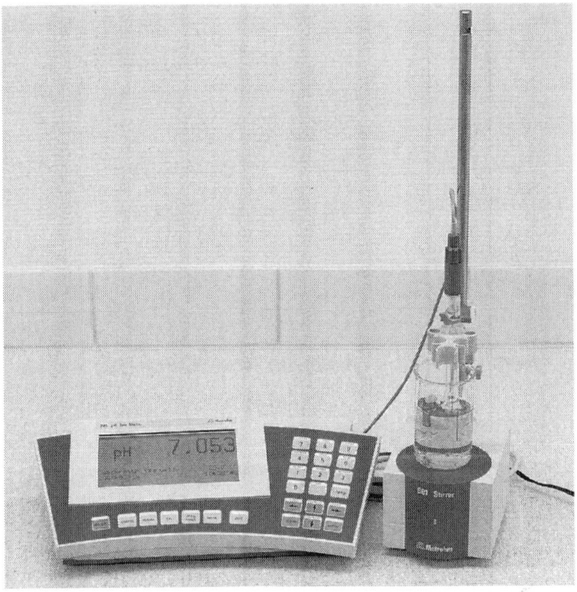

Figure 13.2: pH analysis in a chemical reaction.

Source: By Datamax - Own work, Public Domain, https://commons.wikimedia.org/w/index.php?curid=4377808

13.1 ACIDS

Because of their aqueous solution characteristics, acids are distinguished from other compounds:

- The flavor is unpleasant because of the acidity.
- Changes the color of the litmus test from blue to red.
- When hydrogen gas is combined with certain metals, it emits hydrogen gas.
- When it reacts with bases, it produces salt and water.

Acidic solutions have a pH of less than 7, with lower values indicating higher acidity. Acids include sulfuric acid (used in vehicle batteries) and tartaric acid (found in vinegar) (used in baking). Acids are commonly classified into one of three groups:

Hydronium ions (H_3O^+) are hydronium ions that may be formed by Arrhenius acid in solution (Karsenti, 2008).

- Brnsted-Lowry acid: any chemical capable of donating protons to another.

Any chemical that can receive a pair of electrons is referred to as a Lewis acid.

13.2 STRONG ACIDS ACCUMULATION IN AN ACIDIC SOLUTION

Acid strength, which is frequently measured in solution, is directly related to proton loss and proton donation. Dissociation happens more quickly in the presence of a stronger acid than the acid itself. The six most frequent and highly reactive acids are as follows:

- It is hydrochloric acid in the form of hydrochloric acid (HCl)
- Together with hydrobromic acid (HBr)
- Iodic acid is a kind of hydroiodic acid (HI)
- A chemical compound is hydrogen chloride (H_2SO_4; only the first proton is considered strongly acidic)
- Nitrate is a nutrient-rich molecule in addition to nitric acid. Yes, it is the proper word (HNO_3)
- Perchlorate is a chemical compound ($HClO_4$)

$$P_4O_{10} + H_2O \rightarrow H_3PO_4$$

	Before	After
P	4	1
O	10+1	4
H	2	3

Figure 13.3: A chemical equation.

Source: By Ephert - Own work, CC BY-SA 4.0, https://commons.wikimedia.org/w/index.php?curid=38751665

In solution, each of these acids virtually completely ionizes. Strong acids, by definition, dissociate fully in water; one mole of HA produces one mole of H^+ and one mole of the conjugate base, A^-, with no protonated acid HA left in solution. A weak acid will only partially dissolve in the solution because its proton is less likely to be released (Patil et al., 2008). At equilibrium, acid and conjugate base, as well as a significant quantity of undissociated HA, will be present.

13.3 OTHER SUBSTANCES AND ACIDITY

Two factors must be considered when estimating an acid's total potency:

- A molecule's polarity is
- The link between H and A's tensile strength

In reality, these two characteristics are inextricably linked. As the polarity of the molecule increases, the electron density moves away from the proton. When a proton has a higher partial positive charge, the H-A bond weakens and the proton dissolves more easily. In many circumstances, acid strength is thought of in terms of conjugate base stability. Stronger acids have greater Ka values and lower pKa values than weaker acids.

13.3.1 Bases

There are three methods to define a base:

- When an Arrhenius base is dissolved in water, a hydroxide ion is produced (OH^-).
- A Bronsted-Lowry base, which is a chemical, may accept protons.
- A Lewis base is any substance that can give an electron pair.
- The pH of basic solutions in water ranges from 7 to 14, with 7 being neutral and 14 being the most basic.

13.4 BOTH BASIS POWER AND BASIS POWER CAN BE UTILIZED SIMULTANEOUSLY

A strong acid, for example, may swiftly donate protons, whereas a base can rapidly deprotonate (i.e. remove an H+ ion) molecules, making it the antithesis of a strong acid. Because the deprotonated species is mainly water, basic aqueous solutions in water are often referred to as acidic solutions (Ogino et al., 2013). Deprotonated water naturally produces hydrogen ions. When the pH rises, the concentration of hydrogen ions increases.

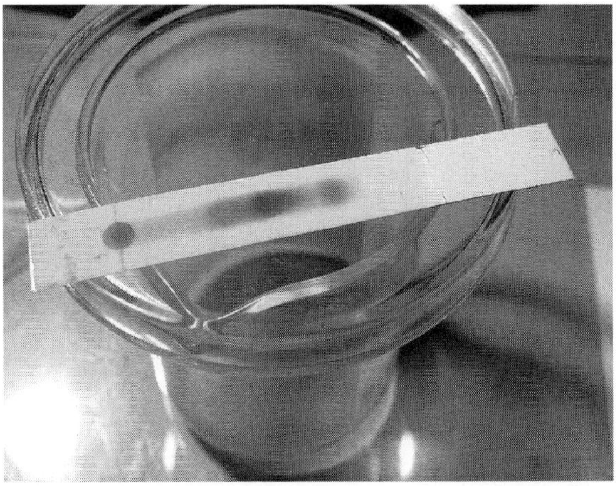

Figure 13.4: An analysis (TLC).

Source: By The original uploader was Natrij at English Wikipedia. - Transferred from en.wikipedia to Commons., CC BY-SA 3.0, https://commons.wikimedia.org/w/index.php?curid=514759

This means that in solution, most alkali metal hydroxides and some alkaline earth minerals can create strong bases. Here are just a few examples:

- Sodium hydroxide, known as (NaOH)
- • Potassium hydroxide solution (KOH)
- Lithium hydroxide (LiOH)
- Rubidium hydroxide (RuOH) (RbOH)
- cesium hydroxide (CsOH)
- Ca (OH) 2: (calcium hydroxide)
- Ba (OH) 2 (barium hydroxide)
- Sstrontium hydroxide

In solution, alkali metal hydroxides dissociate. Although alkaline earth metal hydroxides are more difficult to dissolve than other bases, they are quite effective.

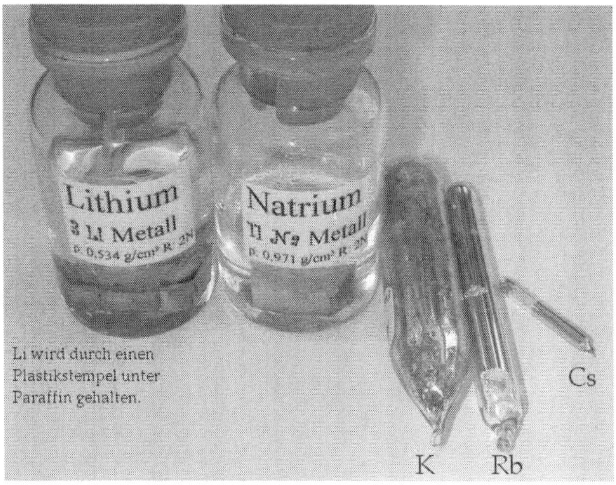

Figure 13.5: Alkali metals.

Source: By Tomihahndorf, CC BY-SA 3.0, https://commons.wikimedia.org/w/index.php?curid=1740883

13.5 WHAT IS THE DISTINCTION BETWEEN ACIDS AND BASES?

An acid is defined as any substance that can donate one of its protons (H^+) to another molecule. Bases, as the name implies, are ions or molecules capable of absorbing a hydrogen ion from an acid. Because of their acidity, acidic substances have a sour flavor (Kawser Hossain et al., 2016). Even after losing its H^+ ion, an acid retains energy benefits over other compounds due to its capacity to provide an H^+ ion. When exposed to acids, blue litmus paper becomes crimson.

Bases, on the other hand, are velvety in texture and have a powerful flavor. An alkali is a base that may be dissolved in water. When these molecules react chemically with acids, salts arise. It is also generally known that when exposed to litmus solution, the bases become red.

13.5.1 Acid-Base Connection Analysis and Interpretation

There are three methods to define acids and bases. Among these hypotheses are the Arrhenius hypothesis, the Bronsted-Lowry theory, and the Lewis theory of acids and bases. In this part, we'll take a short look at each of them. Acids and bases can be described using one of the three principles listed below:

In acidic solutions, there are two types of H^+ ions: those produced by acid and those produced by an alkaline solution. According to Bronsted and Lowry, an acid is a proton donor, and a base is an acceptor. For acids and bases, there exist "electron acceptors" and "electron donors," as described by Lewis.

13.5.2 The Theory of Acids and Bases by Arrhenius

According to the Arrhenius hypothesis, the H^+ ion is generated when acid is dissolved in water. The solution's H^+ ion concentration increases. When the base is dissolved in water, it ionizes the OH^- ion. As a result, the solution contains a lot of OH^- ions. Arrhenius acid can be used to enhance the concentration of protons or H^+ ions in an aqueous solution (Kharat et al., 2017).

Figure 13.6: Nitric acid in a chemical bottle.

Source: https://media.istockphoto.com/photos/nitric-acid-in-glass-chemical-in-the-laboratory-picture-id1350071266?k=20&m=1350071266&s=612x612&w=0&h=pqe5b9pVKlFszNzjoPUGXnIXMeIMCUK_qzXKRcnPq5k=

13.5.3 Arrhenius Acids

Svante Arrhenius, a Swedish scientist, proposed the Arrhenius theory of acids and bases in 1884. According to his concept, acids and bases should be categorized based on the type of ions they produce when dissolved in water (O'Gorman et al., 2012). For example, a hydrochloric acid solution in water. To summarize, HCl degrades into H^+ and Cl^-. The production of hydronium ions raises the concentration of H^+ ions.

H+ (aq) + aqueous HCl (aq)

In the lab, HCl (aq) + H_2O (l) = H_3O+ (aq) + Cl^- (aq)

13.5.4 Arrhenius base

To increase the concentration of water-soluble hydroxide ions in a solution, substances such as an Arrhenius base are utilized. The material must be both water-soluble and capable of dissociating into sodium and hydroxide ions to be classified as an Arrhenius base. When NaOH completely dissolves in water, it produces hydroxide ions and sodium ions, which are used to raise the concentration of hydroxide ions in an aqueous solution.

There is OH^-, Na^+ (aq), and NaOH (aq).

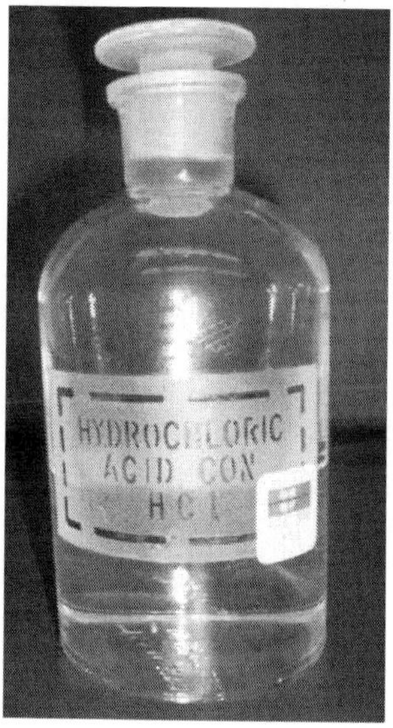

Figure 13.7: Hydrochloric acid.

Source: Public Domain, https://commons.wikimedia.org/w/index.php?curid=95447

13.6 BRONSTED'S THEORY OF ACIDS AND BASES

Bronsted Lowry Acid is named after its creator, Bronsted Lowry. According to the Bronsted-Lowry theory, acids are proton givers. According to this theory, bases are proton acceptors (or H+ ion acceptors). When Bronsted acids degrade and release protons, there are more H^+ ions in the solution. Combine basic acid with a proton.

13.6.1 Bronsted Lowry's bases

Bronsted bases, on the other hand, produce hydroxide ions by absorbing photons from the solvent's water. Bronsted-idea to precisely characterize the acidic or basic characteristics of ionic species, Lowry's of acids and bases can be employed (Neese, 2012). This theory is flawed since it fails to account for acidic non-hydrogenated chemicals such as BF3 and AlCl3.

An acid conjugate (basic and protonic)

Strong Bronsted acids are Bronsted-Lowry acids with a strong proclivity for proton donation but weak conjugate bases. The conjugate base of weak acids is less likely to supply a proton than the conjugate base of strong Bronsted-Lowry acids.

Lewis' Acids and Bases Theory

Lewis defined a Lewis acid as a species with an empty orbital that can accept an electron pair. A Lewis base can act as an electron donor since it has just one pair of electrons. As a result, acids and bases no longer include a hydrogen atom. Lewis acids and bases have two properties: electrophilicity and nucleophilicity.

Lewis acids include Cu2+, BF3, and Fe3+, to name a few. F-, NH3, and C2H4 are the three Lewis bases (ethylene).

A coordinated covalent bond is formed when an electron pair from a Lewis base accepts an electron pair from a Lewis acid. The final product of this reaction is a Lewis adduct. Using this concept, many chemicals may be classed as acids or bases. This isn't the ideal site for individuals who want to learn more about acid-base strengths. Acid-base interactions do not result in covalent coordination.

13.7 PH VALUES FOR ACIDS AND BASES

The pH scale can be used to determine a substance's acidity or basicity (pH stands for "potential of hydrogen"). The pH scale is the most popular and accurate technique to evaluate whether a substance is acidic or basic. There are 14 pH scale points in all, with 0 indicating the most acidic solution and 14 representing the most basic solution. A substance's acidity or alkalinity can be determined using litmus paper (Krylov & Gill, 2013). Litmus paper can distinguish between acids and bases. Today's market offers both red and blue litmus paper. When put in an alkaline or basic solution, blue litmus paper becomes scarlet, indicating the presence of acids or bases.

Figure 13.8: Acid base test.

Source: By Bordercolliez - Own work, CC0, https://commons.wikimedia.org/w/index.php?curid=16884915

It is critical to understand the distinctions between acids and bases.

- Acids have a wide range of characteristics.
- Acids are great conductors of electricity.
- Their caustic qualities make them unsuitable.
- Metals and acidic chemicals mix to generate hydrogen gas.
- As a result, their pH levels are frequently lower than 7 and so, acids have a sour flavor.

Acids include hydrochloric acid (HCl) sulfuric acid (H_2SO_4), and acetic acid, to name a few (CH_3COOH). These are the necessary elements. All bases' bitterness may be tasted. Furthermore, the bases are an excellent complement to the set. Soapy when touched and conducts electricity when submerged in water due to the presence of charged particles in solution are two factors that contribute to its slickness (Kleber & Johnson, 2010). When these chemicals are dissolved in water, they produce hydroxide ions (OH-ions) (OH-ions). Because bases dissolve quickly in water and form a solution, they are excellent electrical conductors. Bases often have a pH value greater than 7. When exposed to acids, blue litmus paper turns red. Because of their disagreeable taste, bases should be avoided.

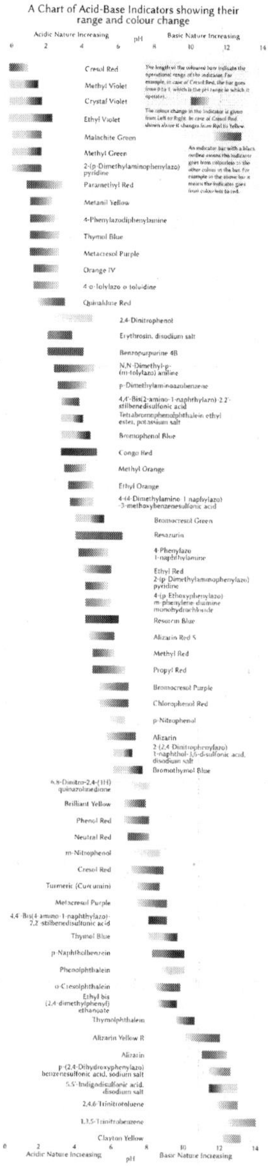

Figure 13.9: A pH scale for measuring alkalinity and acidity of chemicals.

Source: By Damitr - Own work, CC BY-SA 4.0, https://commons.wikimedia.org/w/index.php?curid=35224038

Milk of magnesia, calcium hydroxide, and sodium hydroxide are examples of NaOH hydroxides. As neutral substances are neither acidic nor basic, these compounds do not affect the color of the litmus paper.

- Neutral substances have no impact on litmus paper that has been colored blue or red (Nwodo et al., 2012).
- Pure water has an alkaline pH of seven, making it a healthy drinking option.
- Water and table salt (NaCl) are two examples of neutral items that fit into this category.

Acids and bases are required in a variety of applications.

Among the numerous applications for acids and bases are the following:

13.8 ACIDIC SUBSTANCES AND THEIR APPLICATIONS

- Acetic acid is diluted in vinegar, which may be used for several household tasks. This component is most commonly used as a preservative.
- Orange and lemon juices have a high concentration of citric acid. It is also used as a preservative in the food business.
- Sulfuric acid is increasingly being utilized in batteries. This acid is commonly used in vehicle starting batteries (Laganà et al., 2019).
- Using nitric and sulfuric acids, pigments, explosives, paints, and fertilizers are manufactured.
- Phosphoric acid is a crucial ingredient in a variety of popular sodas.

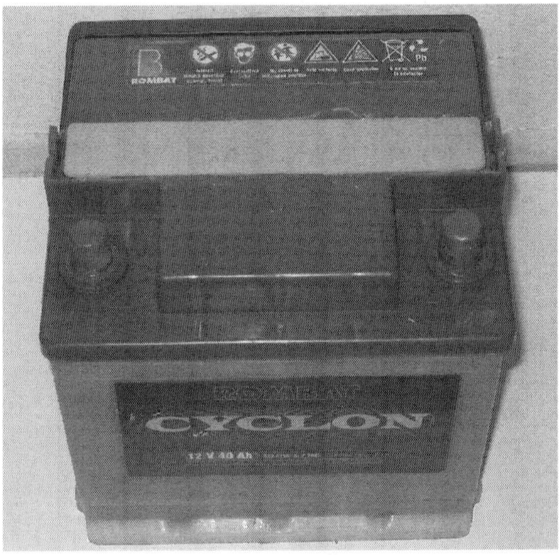

Figure 13.10: Lead acid battery.

Source: By No machine-readable author provided. Shaddack assumed (based on copyright claims). - No machine-readable source provided. Own work assumed (based on copyright claims)., Public Domain, https://commons.wikimedia.org/w/index.php?curid=401224

13.9 APPLICATIONS BASED ON THE FOUNDATIONS

- Sodium hydroxide is used in the manufacture of soap as well as paper. Sodium hydroxide (NaOH) is also used in the manufacture of rayon (NaOH).
- The principal component of bleaching powder is calcium hydroxide (Ca (OH) $_2$, also known as slaked lime or calcium hydroxide.
- It is possible to use dabs of calcium hydroxide dry paint or an aesthetic feature.
- Magnesium milk is an efficient constipation treatment (magnesium hydroxide). As an antacid, it reduces the amount of acid in the stomach.
- Ammonium hydroxide is a frequent reagent in laboratories.

By adding hydrated lime into the soil, the acidity can be reduced. Acids and bases are two of the most fundamental chemical kinds to comprehend chemical interactions. Lemons and cheese contain both citric acid and lactic acid. Ammonia and chlorine-containing bleach are often used as cleaning agents (Lisio et al., 2019). Arrhenius, Bronsted, and Lewis contributed much to our understanding of acid and base. Because of their amphoteric characteristics, they may be both acids and bases. Water is the most well-known example of amphoteric nature since it can function as both an acid and a base.

13.10 WHAT EXACTLY DOES THE TERM "ACID" MEAN?

This assertion is true for both water and aqueous solutions. Water is made up of hydrogen ions and hydroxide ions and is capable of neutralizing acids and bases. This bondt limits the categorization process. The interaction between OH^- and H^+ produces a radical. It has an asterisk next to it (1) Hydrogen peroxide (H_2O) is a chemical compound (1)

According to Bronsted's hypothesis, a substance is acidic if a chemical reaction produces a proton (Hydrogen ion donor). However, Bronsted went on to elaborate on what he had said. Because it is composed of acids, amines, and alcohols, it generates H+ ions. Bronsted-definition In general, Lowry's acid is thought to be the most dependable.

An acidic substance, according to Lewis's theory, absorbs electrons from the other reactant. Electron pairs inhabit the valence shells of acids. The acid oxidation state of the process remains unchanged.

13.11 PHYSICAL QUALITY OF AN ACID

- Acids are distinguished by the following characteristics:
- An electrolyte is a type of molecule that can conduct electricity
- Blue litmus has turned red.
- The vast majority of acids available are solid acids.
- The taste is sour.
- These chemicals react with active metals such as zinc, magnesium, aluminum, and iron to produce hydrogen.
- The pH of acidic surroundings is less than seven.
- A salt and a water solution are formed when an acid reacts with alkaline or basic substance.

- Nature's chemical constitution is primarily composed of acidic chemicals.
- Corrosion can be an issue with some acids.
- Stronger, less volatile acids can be produced more efficiently using salts of weaker, more volatile acids.

13.12 ACIDS AND COMPOUNDS

Acids are classified into two types: those that dissociate ions in water and those that only partially do so. There are two types of acids based on ion dissociation in water: strong acids and weak acids. A strong acid dissolves all of the ions in water. Examples of strong acids include:

The acid HCL (hydrochloric acid)

- HNO3 (nitric acid).
- The hydrobromic acid HBr (hydrobromic acid)

Weak acids partly dissociate ions in water, which is why they are so named. The majority of the acid family is made up of only a few strong acids. When hydrogen and halogen combine to make a weak acid, hydrofluoric acid is formed (Moseti et al., 2016). Despite its classified as a weak acid, hydrochloric acid is exceedingly powerful, corrosive, and strong.

13.13 WHAT EXACTLY DOES THE TERM "BASE" MEAN IN THIS CONTEXT?

In aqueous solutions, a base can transfer electrons, receive protons, or release hydroxide (OH-) ions. A multitude of criteria can be used to identify the base. Bases have a bitter taste, are slick to the touch (e.g., soap), combine with acids to form salts, and accelerate some processes. Many scientists have proposed Arrhenius bases in addition to the Bronsted-Lowry and Lewis bases. Alkaline metal hydroxides, alkaline earth metal hydroxides, and soap are examples of bases.

13.14 CHARACTERISTICS REQUIRED FOR SURVIVAL

Here are some of the benefits of having a foundation:

A basic solution has a pH greater than 7.

Avoid tasting bases since they have a strong flavor.

To create electricity, aqueous or molten base solutions are isolated from one another.

- Caustic bases react negatively with acids and organic molecules.
- pH indicators work as predicted with bases. In the presence of a strong base, bromothymol blue remains blue, but litmus paper becomes blue and yellow in the presence of methyl orange and phenolphthalein pink (Milo et al., 2010).

13.15 DIFFERENT TYPES OF FOUNDATIONS

Bases are available in a range of forms and sizes, as seen below.

- A strong base may accept a proton (H^+) from a weak acid and entirely disintegrate it into its ions in water or another chemical. Strong bases may be made from sodium hydroxide and potassium hydroxide (KOH).
- Weak Base - This sort of base entirely dissolves in water and leaves no trace. A conjugate acid and base are examples of an aqueous solution.
- This molecule forms a covalent bond with a neutral acid.
- The fourth and last Super Base. Because it is made up of an alkali metal and its conjugate acid, this sort of base is more powerful than a strong base. The phrase "super base" (NaH) is used to describe sodium hydride (NaH).
- Its solid base benefits gaseous acids and anion exchange resins. The solid-state is the place to be. Solid bases such as silicon dioxide and sodium hydroxide are used to cover alumina (Metz et al., 2014).

13.16 INTERACTION BETWEEN ACID AND BASE

When an acid and a base combine, a chemical reaction known as an acid-base reaction occurs. This method may be used to calculate the pH of a solution. Several acid-base theories provide distinct viewpoints on reaction processes and how they might be used to tackle associated challenges. One such example is the Bronsted–Lowry acid-base theory.

When the acidic or basic nature of a gaseous or liquid species is not immediately observable, the usefulness of acid-base interactions becomes obvious. Take the acid-base reaction models as a collection of assumptions.

The Bronsted-Lowry theory is a subset of the current Lewis model, which gives the most detailed explanation of acids and bases. In contrast, the Arrhenius hypothesis is the most restricted.

They are unaccountable in any manner. Because a weak base solution is a buffer solution, it is difficult to measure the addition of acid to it. Buffering can also be accomplished with a slightly acidic solution, such as citric acid (Mendoza & Núñez, 2009). When a weak acid and a weak base are mixed, an equilibrium mixture is generated. HPO_2, a hydrogen phosphate ion, can react with adenine, which has the symbol AH.

We may calculate the reaction's equilibrium constant by using the acid dissociation constants of adenine and dihydrogen phosphate ions. The meaning of the acronym [X] is "concentration of X." Subtracting the concentration of hydrogen ions from these two equations yields the equilibrium constant, K.

The flavor of the acids is sour. When acids react with certain metals, hydrogen gas is produced, which can cause corrosion. Citrus fruits (citric acid), vinegar (acetic acid), aspirin (acetylsalicylic acid), automobile batteries (sulfuric acid), and acid rain (sulfuric acid and nitric acid) are all acid-containing items. Although many biological molecules are acids or bases, their structures are more complex and difficult to differentiate. Acids are dietary lipids, whereas purines and pyrimidines are DNA bases. The ability of an acid or basic to ionize determines its strength. Strong acids and bases are completely distinct (Lovley, 2003). When an aqueous solution molecule reaches equilibrium, no trace of the original molecule remains. Weak acids and bases reach an equilibrium state in which most molecules retain their original form. In our daily lives, we come across many examples of acids and bases. Vinegar (acetic acid), lemon juice (citric and ascorbic acid), and wine (which contains tartaric acid) are examples of acid-containing household items (contains tartaric acid). The flavor of these acids is often described as sour. Acids such as hydrochloric acid, sulfuric acid, and nitric acid are more commonly found in laboratories and industries.

The stomach's gastric juices include hydrochloric acid as well. Tannic acid may be found in tea and wine, whereas carbonic acid can be found in carbonated beverages. Acids are also used in the acid etching art form. The process of covering the metal with an acid-resistant waxy material is known as acid etching (Mansour et al., 2010). After exposing the naked metal in a suitable pattern, the sample is steeped in acid. The exposed metal's top layers are permanently removed, resulting in the desired image.

Bases such as sodium hydroxide (also known as caustic soda), ammonium hydroxide, and ammonia are probably familiar to you. Some of these can be found in cleaning products. Bases are often unpleasant in flavor and slick to the touch (a good example being soap) (soap is a good example). Acids and bases are also required commercially in the fertilizer, plastics, and petroleum refining sectors.

13.17 WHAT'S THE DIFFERENCE BETWEEN ACIDS AND BASES?

Acids are chemical substances that, when exposed to water, have a sour taste. They turn blue litmus red. Bases, on the other hand, are chemical substances with a strong flavor and a smooth texture. Some bases dissolve in water, whereas others do not. Alkalis are bases that are soluble in water. They turn red litmus blue. Salts form when acids and bases react with a variety of chemical compounds.

13.17.1 Physical Characteristics

- The term "acid" comes from the Latin word "acidus," which means "sour." This distinguishing feature aids in differentiating acids from other compounds such as salt and bases. Many acids are toxic and should never be eaten (Marian, 2012).
- The acid turns neutral after binding to the base. Water and salt are frequently produced as a result of this interaction. This is a common occurrence when strong acids react with strong bases.
- The acids function as electricity producers. Aside from that, batteries are used to generate electricity. A strong acid effectively conducts electricity, whereas a weak acid inefficiently conducts electricity.

13.17.2 Acids and Bases Have Different Chemical Properties.

1. Acid and Base Metal Reactions

When a metal reacts with an acid, the hydrogen in the acid is often displaced. Hydrogen gas is produced as a result of this procedure. The metals combine with the residual acids to form a salt (Matranga & Corsi, 2012). When the metal interacts with alkalis (water-soluble bases), salt and hydrogen gas are produced (water-soluble bases).

2. Reactions of Acid-Metal Carbonate/Metal Bicarbonate

Salt, carbon dioxide, and water are generated when metal carbonates/ metal bicarbonates react with acids.

3. Reactions of Acids with Metal Oxides

Metal oxides react with acids to produce salt and water.

4. Nonmetal Oxide Reactions with Bases

Salt and water are produced when nonmetal oxides react with bases.

5. Acid and base reactions

When acids and bases react, salt and water are formed. Neutralization reactions are the processes that occur when acids and bases combine to generate salts.

13.18 ACID AND BASE NEUTRALIZATION

The reaction between an acid and a base always results in the formation of salt and water, a process known as neutralization. One H^+ ion of acid is neutralized by one OH^- ion of the base during the neutralization process. Full neutralization happens when all of the H+ ions in the acidic solution are neutralized by the corresponding amount of OH^- ions in the basic solution (Vasapollo et al., 2011). The total number of H^+ and OH^- ions produced by each acid and base determines the proportionate amounts of acid and base required for complete neutralization.

Acids are ion donors of hydrogen. They produce H^+ ions (s) when they react (s).

Acids are neutralized by using bases. They are acceptors of hydrogen I. They get H^+ions when they react. Bases are a subset of alkalis. They contain the OH^-ion and are water-soluble.

In water, strong acids are completely dissociated. This means that the HCl molecules in hydrochloric acid (HCl) completely degrade into $H^+$and Cl^- ions. There is no longer any HCl present.

Water partially dissociates weak acids. This means that some of the CH_3COOH molecules in ethanoic acid (CH_3COOH) degrade to form CH_3COO- and H+, while others remain as CH_3COOH. It has nothing to do with the CH_3 bit being unchanged because the hydrogen is never released. What matters is that some chemicals that create hydrogen atoms can be released as ions, whilst others cannot.

Strong and weak are not synonymous with concentrated and dilute. Concentrated indicates a lot of material in a small amount of solvent, whereas dilute means a lot of solvent in a small amount of substance. Due to the structure of the ethanoic acid molecules, ethanol is generally a weak acid; nonetheless, it is accessible in both dilute (like vinegar) and concentrated (very smelly) forms (Turney et al., 2012). When someone asks if you want your coffee strong or mild, be wary. You can inform them that you can't change the strength, but you can dilute or concentrate it.

13.19 SELECTING AN INDICATOR

We use an indicator to calculate the volume of an acid that only interacts with a certain amount of alkali. We're looking for the volumes of each that result in an equal number of moles of $H^+(aq)$ and $OH^-(aq)$ ions.

A powerful acid paired with a strong base

Assume the alkali and acid have already been put into the flask. The pH begins high (about 14) and remains there until near the end when it drops significantly to roughly pH 1. For this reaction, any indicator that changes color between these pH values would function.

When a strong base and a mild acid are combined,

Assume the alkali and acid have already been put into the flask. The pH begins high (about 14) and remains there until just before completion when it drops rapidly to around pH 8. After that, the pH gradually drops to around 3 (Thomford et al., 2018). For this reaction, any indicator that changes color between 14 and 8 will suffice. Phenolphthalein is a great example (pink at pH > 10 and colorless at pH 10).

When a strong acid and a weak base are mixed,

Assume the alkali and acid have already been put into the flask. The pH starts quite high (about 11) and quickly reduces towards the end when it drops dramatically from around pH 6 to around pH 1. It then continues at a low level. For this reaction, any indicator that changes color between 6 and 1 will suffice. Methyl orange is a good example (red at pH > 4 and orange at pH 4). When a weak acid reacts with a weak base, a weak acid is formed. There isn›t a single reliable signal.

CHAPTER 14

ELECTROCHEMISTRY AND ITS APPLICATIONS

CONTENTS

In one of two ways, chemical reactions can absorb or release electricity as energy. Electrochemistry, a branch of chemistry that investigates the process, is concerned with the conversion of chemical and electrical energy. Electrochemistry has a wide range of practical applications in everyday life. Batteries, which are commonly used in flashlights, calculators, and vehicles, generate energy through chemical reactions (Shin et al., 2019). Electricity is used to apply metals such as gold and chromium to objects. Electrochemistry plays a significant role in the propagation of nerve impulses in biological systems. Redox chemistry, or the movement of electrons, underpins all electrochemical reactions.

Within an electrochemical cell, electricity and chemical energy may be converted back and forth. The following are the three elements of an electrochemical reaction. Only in the presence of a solution that allows for redox reactions can they occur. Conducting these reactions in water facilitates electron and ion mobility. Electron transfer necessitates the use of a conductor. Electrons may go via this conductor, which is often a wire. The salt bridge must be capable of transferring ions through it as part of its function.

14.1 THE ELECTROCHEMISTRY TECHNIQUE

The characteristics of the negatively charged electron impact the interactions between matter and electric current. Atoms, groups of molecules, and the like have a strong attraction to positively charged matter particles like protons. This affinity is similar to the chemical attraction between atoms. Because unbound electrons can either unite with matter particles (reduction) or be liberated by those particles, a change in atoms' electron configurations lies at the heart of all chemical reactions (oxidation). Faraday's equations establish quantitative correlations between free electrons in an electric current and the particles of the material with which they react. Electrolytes or ionic conductors are compounds that engage in electrochemical reactions.

Metals are the most common electrical conductors, and electrons are only available in large proportions in a moderately mobile (free) state. Every electrochemical operation requires the presence of an electron conductor. Because of the energy distribution in the conductor, which restricts electron availability, electrochemical reactions can only occur on the surface of an electronic conductor, a few angstroms away from the conductor. As a result, it is widely assumed that these reactions occur at the interface or common border of an electronic conductor, such as an electrode, and an ionic conductor of electricity, such as an electrolyte (Antoni et al., 2015). In any event, electrochemical processes can only occur in a system that contains electricity. To do this, the system must produce a closed-loop electrically.

Because electrons leave the metal at one metal-solution interface, a lower solution component must exist at a second metal-solution interface where electrons enter the metal during the oxidation process. In conclusion. As previously stated, an electrochemical cell is made up of two electrodes and an ionic conductor (such as an aqueous solution of an ionized salt). Throughout the cell, a redox reaction occurs, with one species being reduced and another being oxidized in different sections of the cell. Electrochemical reactions at electrodes are proportional to current density (expressed in amperes per square centimeter of electrode surface), that is, the current flowing through the cell divided by the electrode surface area (expressed in gram moles per second per square centimeter of electrode surface). Faraday's law states that

14.2 TYPES OF ELECTROCHEMICAL REACTIONS

Electrochemical reactions are classified into numerous categories.

Redox interactions are easy to grasp.

The removal (reduction) or addition (electrolysis) of an electron from the electrode by the charge carrier changes the electrical charge of a charge carrier, which is often a simple or complex ion in solution (oxidation). The same carrier can be found in two distinct states of charge in a solution. The oxidized state has a larger positive charge than the reduced state, which has a smaller one (Shi et al., 2010). An electrode develops a well-defined potential, or reversible redox potential, due to the quickness with which electrons are exchanged with it when ferric and ferrous ions are present in high amounts in solution.

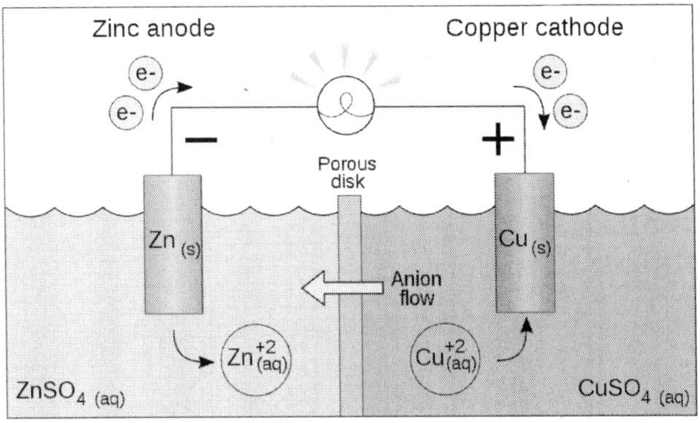

Figure 14.1: Redox reactions.

Source: By Original: Ohiostandard Vector: AntiCompositeNumber - Own work, CC BY-SA 4.0, https://commons.wikimedia.org/w/index.php?curid=84049705

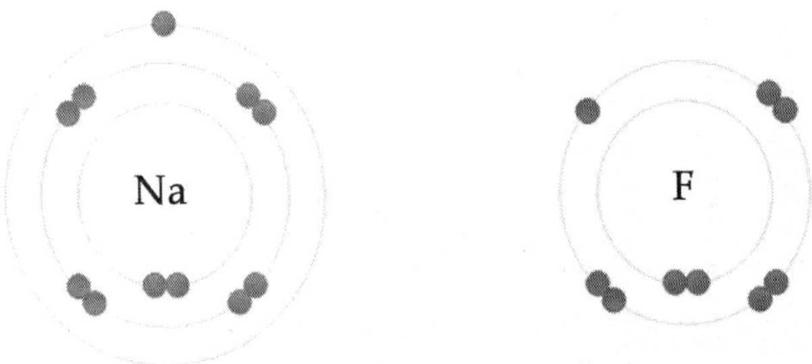

Figure 14.2: Redox reaction with electron transfer.

Source: https://upload.wikimedia.org/wikipedia/commons/a/a8/NaF.gif

14.3 GAS-PRODUCING PROCESSES CREATE GASES AS A BY-PRODUCT

When hydrogen ions in solution react with electrons released from the surface of a metal, they combine with more hydrogen ions and electrons to form gaseous hydrogen molecules. When the reactions between hydrogen ions and hydrogen gas occur rapidly enough, equilibrium is reached. The hydrogen scale, which considers the reversible hydrogen electrode as a metal in contact with a solution and whose electrical potential has been arbitrarily fixed at zero, may be used to measure electrode potentials (Aramwit et al.,

2010). In a similar fashion to that described above, OH ions in solution can be induced to donate electrons to metal, resulting in a series of events that creates oxygen gas. As a by-product of the electrochemical oxidation of chloride ions in concentrated neutral and acidic salt solutions, chlorine gas is generated.

14.4 CHEMICAL REACTIONS TO METAL DEPOSITION AND DISSOLUTION

The electrode's metal lattice collects metal ions that have been reduced to a neutral atom or species and then discharged. In this way, metals may be deposited at electrodes. If positive potentials are given to the metal electrode to remove electrons, the produced metal ions can penetrate the double layer of electric charge at the interface, hydrate (combine with water), and enter the solution. As a result, the metal electrode disintegrates. When metals and their ions interact in solution, they produce these well-defined electric potentials.

14.5 ORGANIC COMPOUNDS CAN BE REDUCED AND OXIDIZED

At electrodes, organic molecules can also be oxidized and reduced. Because these molecules are exceedingly difficult to recombine, they cannot be restored to their former condition in most situations. Quinones, amines, and nitrous compounds (all of which include oxygen and nitrogen) are all known to have reversible potentials.

14.6 SYSTEM OF CHARGE TRANSFER

We can trace electrode reactions' thermodynamically irreversible features to the fundamental nature of charge transfer. In this reaction, like in many chemical processes, an energy barrier occurs between the oxidized and reduced states. Unless a molecule acquires an excited state with an abnormally high energy content via interactions with other species (atoms, ions, molecules, etc.) in its close neighborhood, the reaction will not occur under normal conditions. When a chemical reaction happens, this energy level must be adequate for the species to pass over into the transition state, the largest energy barrier, before it does so in most instances. Calculations reveal that electron exchange activities at electrodes would be unacceptably sluggish if this approach is applied to electron transport at an interface. This contradicts reported instances.

However, electron mobility is regulated by quantum mechanical principles, and these laws are incorporated into the computations to match reality. A hydrated ion in the outer Helmholtz plane must have well-defined quantized energy levels equivalent to those of free electrons in the metal for electron exchange to occur fast in quantum physics (Sneskov & Christiansen, 2012). Electron tunneling over the energy barrier, according to the classical perspective, refers to the rapid electron exchange between an electrode and a particle in solution since the particle can acquire such states with less energy than is necessary for its transmission over the top of the barrier.

Electrochemical processes can be speeded up or slowed down by changing the electrode potential. As the electrode's negative charge increases, so do the rate at which electrons in the metal are ready to tunnel to ions. More electrons are willing to give up electrons if there is a larger positive potential, so the oxidation process happens faster.

For a given concentration of the reactive species, the rate of reaction and electrode potential must be proportional in a linear form.

Both reductions of one species and oxidation of the reduction product occur at each electrode potential, but at different rates; the rates of each reaction are dictated by their respective concentrations and the attendant potential effects. The electric current density, a measure of the number of electrons moving in a given volume of space at a given time, is the most basic method to characterize the rate of an electrochemical process.

Various elements and constants contribute to the net current, which is comparable to the difference in rates of reduction and oxidation.

14.7 DETAILED ELECTROCHEMICAL PROCESSES

For example, the iron ion with two positive charges, Fe++, is an example of a simple electrochemical process in which a single electron is added to a particle to generate a reduced ion (e.g., Fe++). An arrangement of the surrounding solvent molecules may be all that distinguishes reactants and products from simple Ionic Redox reactions. A reaction is typically composed of a series of elementary acts or unit steps called a reaction mechanism when structural changes accompany one or more electron transfers between the electrode and a species in solution (such as when the hydroxyl ions, OH, are transformed into an oxygen molecule, O2, and a water molecule during the oxygen evolution process at the anode, or positive electrode). In most situations, unstable species with a larger energy content than the reactants or products are found in intermediate states between phases (Stuppia et al., 2012).

Electron transfer stages and chemical reactions occuring before, after, or in between electron transfer steps are typical in complex reaction processes. The great majority of organic electrochemical processes need a large number of electrons, making them challenging to explain. Most reactions have a considerable energy barrier that is met at some point in time. The pace at which this phase happens dictates the overall reaction rate (i.e., all other processes must proceed at the same net rate, although they might provide for a considerably quicker total change) (i.e., all other steps must occur at the same net rate, although they could provide for a much faster overall change). Known as the rate-determining step, all previous and succeeding steps can be assumed to be in equilibrium before and following this moment. Butler-Volmer equations apply whenever this is the case, but the transfer coefficients a and c must have unique reaction mechanism-specific values.

14.8 INTRICATE REACTIONS OF NUMEROUS SORTS

14.8.1 Electrocatalysis

Electrocatalysis is concerned with problems pertinent to the acceleration of electrochemical processes, or, to put it another way, the reduction of the overpotential necessary to complete reactions at a given rate. Both gains and decreases are significant because they impact the economics of electrochemical activities. Electrocatalysis is the study of the electrode as a substrate or base for electrochemical processes, in addition to the influence of its bulk and surface features on reaction rate. Contrary to traditional perception, the initial stage of an electrochemical reaction, electron transfer, is unaffected by the ease with which electrons are liberated from the metal. Therefore, temperatures of activation are generally independent of the metal substrate in fundamental electrode operations requiring just electron transport.

In a process involving the establishment of a chemical bond between the electrode substrate and one of the radicals (charged particles) generated on the surface, the reaction rates at a given voltage for varied substrates might vary by several orders of magnitude. When the substrates are transition metals, reliable links between reaction rate and specific properties of the metal's electronic structure are established, and the variation is a consequence of the strength (or energy) of bonds produced between intermediate species and the surface (Rhoads et al., 2012). If these bonds are too weak, the catalytic effect of the substrate is minimal. If they are sufficiently powerful, the intermediate species become too stable to continue interacting, so preventing future reactions on the surface of the metal. If the pace of a reaction is plotted against a feature of the metal substrate that is proportional to the surface's capacity to form bonds, such as its sublimation temperature, the ideal bond strength may be established.

In addition to catalyzing or inhibiting electrochemical processes, foreign species in the electrolytic solution can also catalyze or inhibit electrochemical reactions. In the most typical scenario, ions could alter reaction rates by modifying the characteristics of the double layer via distinct adsorption effects. Organic molecules hinder an electrochemical process by blocking its surface. Some organic ions or dipoles can have a blocking effect at a given voltage. Before reaching equilibrium, the electrode reaction exhibits the typical current density-potential link, which is subsequently suddenly interrupted by an abnormal drop in current, which corresponds to adsorption.

14.8.2 Electro crystallization

Deposition of metals and other compounds at electrodes as a consequence of an electrode process exhibits several different features. The electrode process is followed by crystal formation, and this results in a continual modification of the electrode surface. This shift, in turn, changes the electrochemical characteristics of the system—the double-layer capacity and the rate constants of the charge transfer processes. Hence, if electrochemical characteristics are to be studied, transitory procedures should be employed that allows measurements to be done before major changes in surface morphology (structure) take place.

Since the two phases, the discharge of ions at the electrode and the integration of the discharged ions into the crystal lattice, are separated in time and space, an intermediate species occurs at the surface, that of relatively loosely bound and freely moving atoms, dubbed adatoms. Since the electrons strive to join the rest in the bulk of the metal, adatoms seem to have a partial charge, lower than that of the elementary positive charge (Rastogi & Sani, 2011). The adatoms, therefore, attract solvent molecules, and the species is partially solvated. This reaction entails treating an adatom as a sort of adsorbed ion, called an addition, which, nonetheless, has already experienced partial discharge.

Because such an intermediate state is feasible, ions need not be reduced to the neutral atomic state at the location of integration into the crystal lattice. Indeed, energy considerations indicate this drop to be unattainable. Instead, the discharge of ions is favored on crystal surfaces. Their migration from the source of discharge to a location on the crystal happens through surface diffusion. When such surface diffusion is inhibited, it can, in some scenarios, reduce the rate of reaction.

Problems of the crystal-building process in electro rystallization are in many respects comparable with those of crystal formation from the gas phase. Crystal building can occur in any one of three unique ways.

Kinks, naturally existing on any metal surface, form a suitable half-lattice position at which an atom is surrounded by one-half of the number of atoms that would surround it in the bulk of the metal; there, adatoms can be successively trapped, and thus the crystal lattice is extended along a crystal edge and further on across the surface. The approach, however, has a bonded capacity for crystal production. A step can move as far as the edge of a crystal, and step-growth would lead to smoothing of the surface to perfection, but then further development would halt.

Mechanisms connected with screw dislocations, or twinning edges, might give for the continual evolution of crystals. The screw dislocation mechanism is rendered practical by a particular imperfection frequently observed in the crystal lattice that may be termed a dislocation coming from a displacement of one atom in the lattice concerning a perfect arrangement (Raies &Bajic, 2016). This shift may then result in the development of a monoatomic edge protruding above the electrode surface at which new atoms can be piled. The stacking causes a twisting of the edge around the base atom as a core, the process creating a spiral growth.

If growth sites are scarce, or if the substrate at which deposition should take place is foreign to the depositing metal, the charge transfer leads to a build-up of the atoms to a concentration significantly bigger than that which can exist there in equilibrium with the crystal lattice. In such a situation, termed supersaturation, aggregation of adatoms to form crystal nuclei is favored. Surface energy requirements demonstrate that, at any degree of supersaturation, nuclei of specified dimensions remain stable and may offer prospects for future growth.

The growth of deposits is governed by the aforementioned mechanism as long as the discharge process supplies enough quantities of adatoms as building material along the complete surface. If the rate of deposition is accelerated, to induce near the surface substantial depletion of ions, uneven deposition will ensue. This is generated by the protrusions of a generally rough metal surface being closer to the bulk of solution than the recessed areas and, thus, obtaining a somewhat quicker supply of the discharging species. Once such a situation is developed, it tends to develop further. The rapidly growing points penetrate ever richer layers of solution, resulting in ever speedier development. Thus, a natural consequence of deposition under a transport-controlled rate is the amplification of the original surface roughness and the appearance of projecting spikes, called dendrites, as deposits.

An opposite process is the smoothing of the original surface defects, which may occur when some foreign species, termed an additive, is present in the solution and adsorbs on the surface and slows the process of discharge. If such molecules are incorporated into a growing deposit, a scenario may evolve in which their travel to recessed sections of the surface becomes slower than to elevated regions (Aidas et al., 2014). As a result, deposition becomes faster at recessed locations than at high ones and leveling of the surface occurs. This strategy has tremendous technological applicability. Metal disintegration can occasionally be managed by a similar, transport-controlled process, in which case a polishing effect is created. Electropolishing has found substantial uses in practice.

14.9 ORGANIC ELECTRODE REACTIONS

A comparatively high number of electrochemical reactions involving organic molecules are recognized. Many chemical organic processes may be made to run electrochemically, a typical benefit being that the velocity of the reaction is conveniently managed by adjusting the voltage. A change of voltage may vary the direction of the reaction and hence a certain product may be tuned in. Polymerization activities at electrodes can be expedited; such a technology is applied as a stage in the synthesis of nylon.

14.10 MULTIELECTRODE SYSTEMS

So far, systems have been explored in which a single electrode process takes effect. In principle, at any electrode potential, all species present in the system fall into two categories: those that are stable, and those that undergo oxidation or reduction. The stable species are those that at the specified electrode potential would not diminish their free energy by giving off or absorbing electrons.

At any potential, there should occur a discharge of all unstable species. Thus, the system may be seen as a multielectrode system, consisting of as many electrodes as there are redox couplings present. The rate at which numerous processes occur, however, can be so greatly varied that typically a single process is by far the prevailing one. Systems in which two electrode processes occur at comparable rates are of substantial value. If two kinds of metal ions are released concurrently, an alloy is generated following crystallization (Pritchard et al., 2019). The characteristics of the alloy would in most circumstances be those indicated by the phase diagram (a plot of the temperature of melting versus the composition of the mixed system) for the ratio of quantities of discharged metals supplied by the rates at which the discharges take place. In varied circumstances, nonequilibrium metal phases are generated imparting distinct features to the alloy.

If a process of metal dissolution (an oxidation process) may function at a pace similar to that of some reduction process on the same metal, a corrosion pair is generated. Thus, zinc submerged in an acid solution tends to provide a potential sufficiently positive for the metal to discharge metal ions into the solution (oxidation). At the same time, it is also sufficiently negative for the reduction of hydrogen ions, found in every aqueous solution, to hydrogen gas. Hence, a spontaneous process of hydrogen evolution and dissolving of the metal will take place. The mixed potential, spontaneously formed by the corroding metal, is obtained by equating equations for the anodic and cathodic currents of the two processes in the corrosion pair.

Corrosion can be avoided in two ways. One is by employing an external source to make the potential of the metal sufficiently negative to put it into the potential area in which the metal is stable, called cathodic protection; and the other is by stimulating by some means the production of a coating on the surface that would halt the process (Pinzi& Rastelli, 2019). Such films could consist of an oxide or a layer of organic molecules that prevent disintegration and hence is named an inhibitor.

14.11 APPLICATIONS

14.11.1 Voltaic (Galvanic) Cells

Galvanic cells, commonly referred to as voltaic cells, are electrochemical cells that create electricity by spontaneous oxidation-reduction events. It is typical to break oxidation-reduction reactions into half-reactions when formulating the equations to assist in balancing the entire equation and emphasize the real chemical processes.

Consider what occurs when copper is immersed in a silver nitrate solution. As soon as the copper metal is introduced, silver metal develops, and copper ions move through the solution. The blue color of the solution on the far right shows the presence of copper ions. The reaction may be separated into two half-reactions. Half-processes separate oxidation and reduction reactions, enabling each to be examined individually.

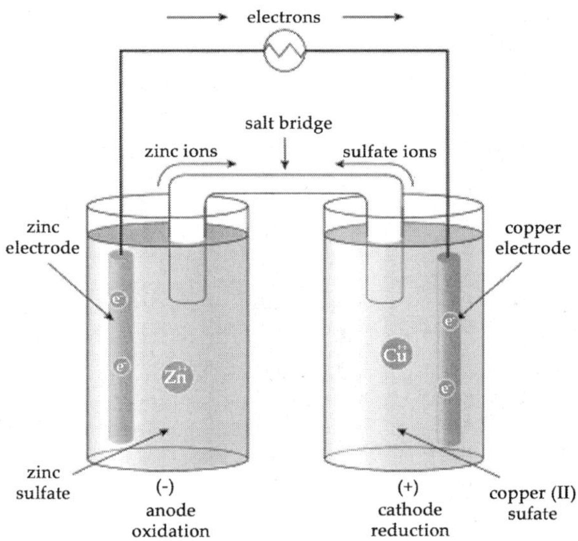

Figure 14.3: A galvanic cell.

Source: https://commons.wikimedia.org/wiki/File:Galvanic_cell_labeled.svg

2(Ag^{+}(aq)+eAg(s))or2Ag+(aq)+2e2Ag+(aq)+2e2Ag+(aq)+2e2Ag+(aq)+2e 2Ag+(aq)+2e2Ag+(aq)+2e2Ag+(aq)+2e2Ag+(aq)+2e2Ag+(aq (s) (s)

Overall: 2Ag^{+}(aq)+Cu(s)⟶2Ag(s)+Cu^{2+}(aq) 2(Ag^{+}(aq)+eAg(s)) or2Ag+(aq)+2e2Ag^{+}(aq)+2e2Ag+(aq)+2e2Ag^{+}(aq)+2e2Ag^{+}(aq)+2e2Ag^{+}(a q)+2e2Ag^{+}(aq)+2e2Ag^{+}(aq)+2e2Ag+(aq (s) (s) 2Ag+(aq)+Cu(s)2Ag(s)+C u2+2Ag(s)+Cu2+ (aq)

The equation for the reduction half-reaction had to be twice so that the number of electrons "gained" in the reduction half-reaction would match the number of electrons "lost" in the oxidation half-reaction.

The copper metal undergoes an oxidation-reduction process when put in a transparent solution of silver nitrate (a), resulting in the exchange of Cu^{2+} for Ag^{+} ions in the solution. Due to the presence of copper ions, the solution becomes blue (c) as the reaction develops (b), and silver metal is deposited on the copper strip as the silver ions are withdrawn from the solution.

In galvanic or voltaic cells, the half-processes of spontaneous electrochemical reactions are separated, enabling current to flow via an external wire. There is a beaker holding a 1 M solution of copper (II) nitrate [Cu $(NO_3)_2$] and a piece of copper metal partially buried in the solution. Copper is an electrode material. Since copper oxidizes, the copper electrode works as the anode. A wire links the anode to a voltmeter, while a second wire connects the opposite end of the voltmeter to a silver electrode. The silver electrode functions as the cathode since it is being decreased (Patil et al., 2008). The silver electrode is submerged in a 1 M solution of silver nitrate in the right-hand half-cell of the figure ($AgNO_3$). Since the circuit is open, there is no current, or considerable movement of electrons via the wire. A salt bridge completes the circuit by carrying electricity through ion movement.

The salt bridge is constructed of a concentrated, nonreactive electrolyte solution, sodium nitrate ($NaNO_3$) in this case. As electrons travel from left to right through the electrode and wire, nitrate ions (anion) migrate through the porous plug on the left into the copper (II) nitrate solution. The charge on the copper (II) ions created in the solution when copper metal is oxidized is neutralized, maintaining the electrical neutrality of the beaker on the left. Sodium ions (cations) migrate to the right, through the porous plug, and into the solution of silver nitrate to the right, whereas nitrate ions move to the left (Ogino et al., 2013). These additional cations "replace" the silver ions that are eliminated when the solution is reduced to silver metal, retaining the electrical neutrality of the beaker on the right. When the electrochemical cell is created in this manner, a positive cell potential (0.46 V) implies a spontaneous reaction and left-to-right electron movement.

- In a normal galvanic cell, electrons travel left to right from the anode to the cathode.
- Since oxidation happens here, the left half-cell electrode is the anode. The term is a reference to the anions that flow toward the salt bridge.
- The cathode electrode in the right half-cell is the cathode since reduction happens there. The term relates to the cation movement toward the salt bridge (Arif et al., 2010).
- The anode is the location of oxidation (the left half-cell in the picture) (the left half-cell in the figure).
- At the cathode, a reduction occurs (the right half-cell in the picture) (the right half-cell in the figure).
- The salt bridge must be present and oxidation and reduction must occur for the current to flow for the circuit to be closed.

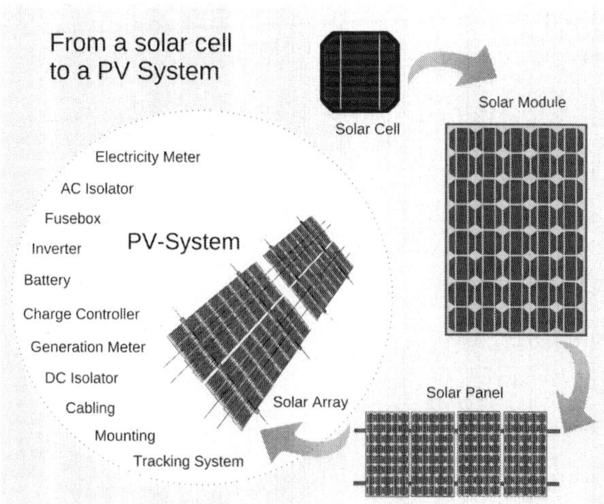

Figure 14.4: Solar panels are also examples of voltaic cells.

Source: By Rfassbind - Own work, Public Domain, https://commons.wikimedia.org/w/index.php?curid=34961018

14.12 BATTERIES

A battery is an electrochemical cell or set of cells that creates an electric current. Theoretically, any galvanic cell may work as a battery. An ideal battery would never lose power, provide consistent voltage, and be able to tolerate extremes in temperature and humidity. Real batteries create a compromise between optimal performance and practical constraints. An automobile battery, for instance, weighs around 18 kilos, or approximately 1 percent of the mass of a standard car or light truck. This sort of battery may give practically endless energy if utilized in a smartphone, however, it would be rejected owing to its size (O'Gorman et al., 2012). As a result, there is no "optimal" battery, and batteries are selected for specific applications depending on their mass, cost, dependability, and current capacity. The two fundamental battery kinds are primary and secondary batteries. Following that, we will explore a few examples of each battery type.

Figure 14.5: Batteries.

Source: By en:User:Brianiac - en:Image:Batteries.jpg, Public Domain, https://commons.wikimedia.org/w/index.php?curid=209635

14.13 DRY CELLS (PRIMARY BATTERIES)

Primary batteries, unlike secondary batteries, cannot be recharged. The dry cell is a popular primary battery option. In the dry cell, a zinc-carbon battery is used. Zinc is used for both the container and the negative electrode. A positive electrode is surrounded by ammonium chloride, manganese (IV) oxide, zinc chloride, carbon powder, and a trace amount of water. The anode reaction can be used to describe the typical zinc oxidation:

$Zn(s) \longrightarrow Zn^{2+}(aq)+2e-$

$Zn(s) \longrightarrow Zn2^{+(}aq)+2e-$

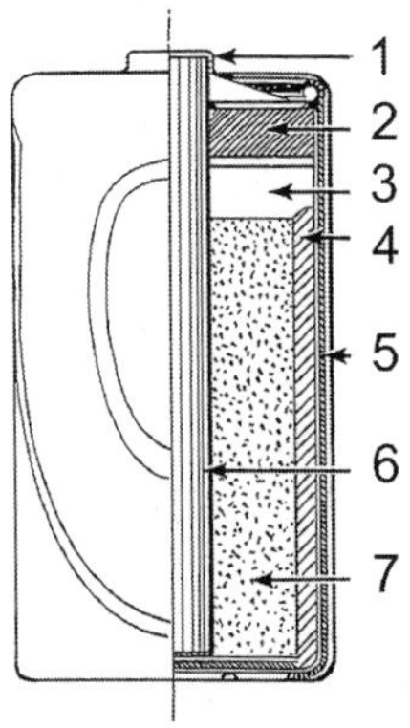

Figure 14.6: A dry cell (1. brass cap, 2. plastic seal, 3. expansion space, 4. porous cardboard, 5. zinc can, 6. carbon rod, 7. chemical mixture).

Source: By Pearson Scott Foresman - This file has been extracted from another file, Public Domain, https://commons.wikimedia.org/w/index.php?curid=6968348

The cathode reaction is more difficult because it involves multiple reactions. The cathode's chain of events is about

$2MnO_2(s)+2NH4Cl(aq)+2e- \longrightarrow Mn_2O_3(s)+2NH_3(aq)+H_2O(l)+2Cl)$

$2MnO_2(s)+2NH4Cl(aq)+2e- \longrightarrow Mn_2O_3(s)+2NH_3(aq)+H_2O(l)+2Cl-$

The following is a summary of the zinc-carbon battery process:

$2MnO_2(s)+2NH_4Cl(aq)+Zn(s) \longrightarrow Zn_{2+(}aq)+Mn_2O_3(s)+2NH_3(aq)+H_2O(l)+2$ Cl–

$2MnO_2(s)+2NH_4Cl(aq)+Zn(s) \longrightarrow Zn^{2+}(aq)+Mn_2O_3(s)+2NH_3(aq)+H_2O(l)+2$ Cl–

With an initial cell potential of about 1.5 volts, this potential decreases as the battery are depleted. A battery delivers the same amount of voltage regardless of its size. Because of this, AA, AAA, and D batteries all have the same voltage rating. Larger batteries, on the other hand, can supply more electrons. Only a small amount of MnO_2MnO_2 in the vicinity of the cathode is reduced in the dry cell during discharge, making it a somewhat inefficient method of generating electrical energy. To make matters worse, the anode of dry cells spontaneously interacts with the electrolyte NH_4ClNH_4Cl, causing the casing to corrode and leak. Alkaline batteries were developed in the 1950s to address some of the issues with zinc-carbon dry cells. These devices can be used to replace zinc-carbon dry cells. As the name implies, alkaline batteries use potassium hydroxide as their electrolyte (Barbatti et al., 2014). An alkaline battery may provide three to five times the energy of a zinc-carbon dry cell of the same size. Alkaline batteries should also be removed from gadgets for long-term storage because they are prone to leaking potassium hydroxide. Some alkaline batteries can be recharged, but the vast majority of them cannot. Attempts to recharge a non-rechargeable alkaline battery usually result in the battery bursting and the potassium hydroxide electrolyte being released into the environment.

Figure 14.7: Primary batteries.

Source: By Lead holder - Own work, CC BY-SA 3.0, https://commons.wikimedia.org/w/index.php?curid=15389714

14.14 LEAD-ACID SECONDARY BATTERIES (SECONDARY BATTERIES)

As a backup battery, your car has a lead-acid battery. Secondary batteries can also be recharged. It is both cheap and efficient to use a lead acid battery to power a car's starter motor. The reactions of a lead-acid battery are

$Pb(s) + H_2SO_4(aq)$ $PbSO_4(s) + H^{2+}(aq)$ in the anode (aq)

$PbO_2(s)+H_2SO(aq)+3H^+(aq)+2e$cathode: $PbO_2(s)+HSO_4(aq)+3H^+(aq)+2e$cathode

$PbSO_4(s)+2H_2O(l)$

The total is $Pb(s)+PbO_2$ (s)

$PbSO_4(s) + 2H_2O(l)$ (l)

$Pb(s)+HSO_4$ in the anode

$PbSO_4(s)+H+(aq)$

When a second electron is present, the cathode is made up of the following compounds:

$PbSO_4(s)+2H_2O(l)$

$Pb(s) + PbO_2(s) + 2H_2SO_4(aq) + 2H_2O(l)$ (l) is the total reaction

Six cells are connected in series to form a 12-volt battery for an automobile. Despite their bulk and corrosive liquid electrolyte, lead-acid batteries are still widely used due to their high current density. The lead-acid battery in your car is made up of six cells that are wired in series to produce 12 volts. These devices are ideal for starting motors in automobiles due to their low cost and high current output (Bacellar et al., 2015).

Unlike a dry cell, a lead storage battery can be recharged. As a result of the forward redox reaction, solid lead (II) sulfate accumulates on the plates over time. The amount of sulfuric acid is also reduced. When the car is driven regularly, the generator recharges the battery by forcing the above-mentioned reactions to go in the opposite, or non-spontaneous, direction from what they would normally go.

$2PbSO_4(s)+2H_2O(l)\rightarrow Pb(s)+PbO_2(s)+4H^+(aq)+2SO^{2-}4(aq$

$2PbSO_4(s)+2H_2O(l)\rightarrow Pb(s)+PbO_2(s)+4H^+(aq)+2SO4^{2-}(aq)$

During this process, lead, lead (IV) oxide, and sulfuric acid are all regenerated. A lead storage battery, as far as we know, has an infinite lifespan. During recharging, lead (II) sulfate falls from electrodes and settles on the bottom of cells, making it less than ideal. Because of the high lead content, lead storage batteries must always be disposed of properly.

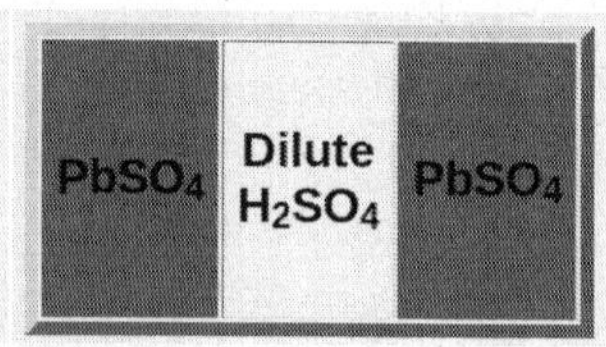

Figure 14.8: Discharging of lead acid batteries.

Source: By Riventree at English Wikipedia, CC BY 3.0, https://commons.wikimedia.org/w/index.php?curid=100305176

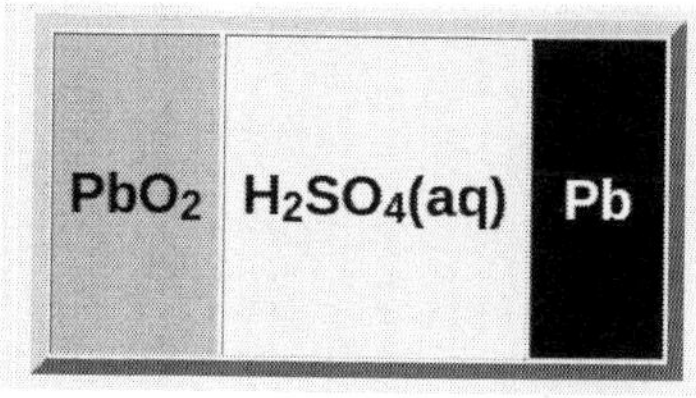

Figure 14.9: Charging in a lead acid battery.

Source: By User:Riventree, CC BY 3.0, https://commons.wikimedia.org/w/index.php?curid=104765945

14.15 OTHER ENERGY SOURCES

Nickel-cadmium, or NiCad, batteries are used in small electrical appliances and equipment such as drills, portable vacuum cleaners, and AM/FM digital tuners. This battery is made up of a nickel-plated cathode, a cadmium-plated anode, and a potassium hydroxide electrode. To prevent shorting between the plates, they are rolled into a case with the separator in place (Neese, 2012). The NiCd cell can produce significantly more current than a comparable alkaline battery of the same weight and dimensions thanks to its "jelly roll" design.

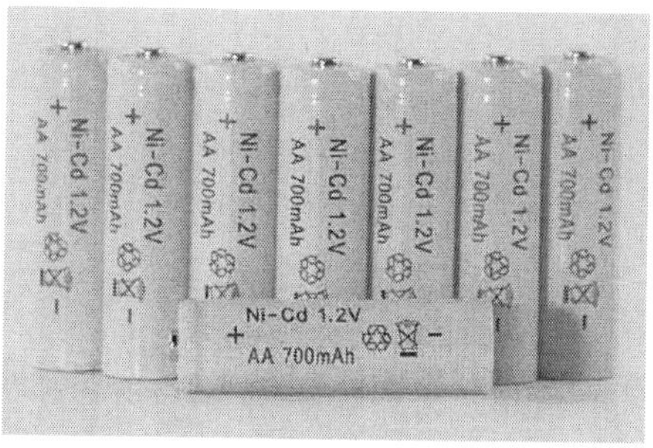

Figure 14.10: Examples of nickel cadmium batteries.

Source: https://www.istockphoto.com/photo/rechargeable-aa-battery-gm91783463-9675664

The voltage drops to between 1.2 and 1.25 volts when the battery is depleted. If handled properly, a NiCd battery can be recharged up to 1000 times. NiCd batteries should never be opened or discarded because cadmium is a toxic heavy metal. The majority of portable electronic devices use lithium-ion batteries, which are among the most widely used rechargeable batteries. The voltage of the battery is approximately 3.7 volts. Lithium batteries have several advantages, including their ability to deliver a significant amount of current, their lightness, and their relatively constant voltage as they drain (Nwodo et al., 2012). Specialized lithium-iodide (polymer) batteries are used in many long-lasting, critical pieces of equipment, such as pacemakers and other surgically implanted electronic medical devices. These devices are designed to last for at least 15 years.

Figure 14.11: Lithium ion battery.

Source: By Kristoferb, CC BY-SA 3.0, https://commons.wikimedia.org/w/index.php?curid=12279456

Keep secondary lithium-ion or lithium-polymer batteries separate from disposable primary lithium batteries. When we say "lithium battery," we're referring to a wide range of lithium-metal chemistries with various anodes and cathodes. Lithium batteries are frequently used in consumer electronics and electric vehicles of all types, ranging from full-size automobiles to radio-controlled toys.

Figure 14.12: Vehicles powered by lithium ion batteries.

Source: https://www.istockphoto.com/photo/inside-electronic-car-battery-pack-rechargeable-cells-inside-chassis-components-3d-gm1330889092-414177892

Single-cell watch batteries (also known as coin or button cells) come in a variety of shapes and sizes, but they all share one feature: they are all made of one cell. A metal can serves as both the negative terminal and the bottom body of the cell. The negative terminal is protected by an insulated top cap. A button cell is a single cell that may have been one of the first to form in the body. The most common anode materials are zinc and lithium (Bennett et al. 2010). The most common cathode materials are carbon monofluoride, cupric oxide, silver oxide, and manganese dioxide. Mercuric oxide button cells were once widely available but are no longer due to mercury toxicity and the environmental consequences. Button cells are particularly dangerous to infants and young children. Button cell burns to the inside of the body can be fatal.

14.16 MOBILE POWER PLANTS (FC)

Electrochemical devices that convert chemical energy into electrical energy are known as fuel cells (FC). Fuel cells, like batteries, require a constant supply of fuel, which is typically hydrogen. They will continue to produce electricity as long as fuel is available. Hydrogen fuel cells have powered satellites, space capsules, automobiles, boats, and submarines.

A voltage of nearly one-hundredth of a volt is present. Hydrogen fuel cells have an efficiency of 40-60%, which is higher than that of a typical internal combustion engine (25-35%), and fuel cells emit only water as exhaust. Fuel cells have been unable to be used in the majority of applications due to their high cost and technological complexity.

14.17 ELECTROLYSIS

This section discusses galvanic cells, which generate electricity through a spontaneous chemical reaction. Electrolysis in an electrolytic cell, as opposed to spontaneous reactions, is the result of an external voltage. Electrolysis has numerous commercial and industrial applications. It is a method based on electrolysis for separating metals from metallic compounds to obtain metals in their purest form. This method can produce aluminum, lithium, sodium, potassium, magnesium, calcium, and, in some cases, copper (Moseti et al., 2016). Brine is electrolyzed to produce chlorine and sodium hydroxide from a mixture of water and sodium chloride. In spacecraft and nuclear submarines, water electrolysis is used to produce oxygen.

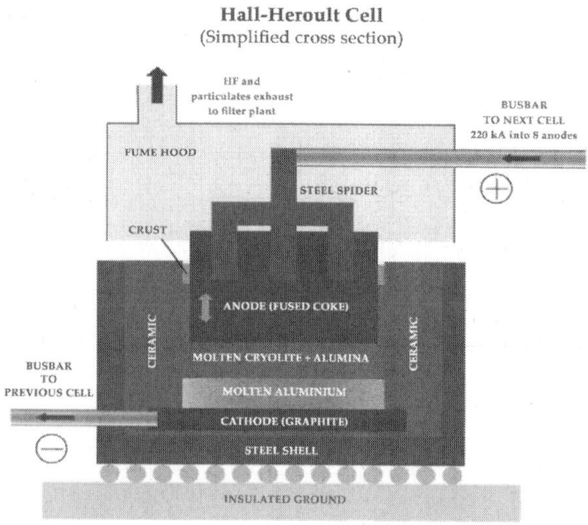

Figure 14.13: Electrolysis.

Source: By Kashkhan at English Wikipedia, CC BY-SA 3.0, https://commons.wikimedia.org/w/index.php?curid=6864903

Cleaning and preserving historical artifacts can also be accomplished through the use of electrolysis. Because the process separates non-metallic particles from metallic particles, it can be used to clean a wide range of metallic objects, from old coins to larger objects like rusted cast iron cylinder blocks and heads when reassembling automobile engines (Milo et al., 2010). Rust can be removed from small iron or steel objects in a home workshop using electrolysis and inexpensive supplies like a plastic bucket, tap water, rebar, washing soda, baling wire, and a battery charger.

It is common practice to use electrolytic cells in ***electroplating***. Electroplating is a process that coats a conducting surface with a thin layer of a specific metal. An object can be made more corrosion-resistant, the surface can be strengthened, or a more appealing finish can be achieved by electroplating. Typical electroplating metals include cadmium, chromium, copper, gold, nickel, silver, and tin. Silver- or gold-plated tableware, chrome-plated automobile parts, and jewelery are among the most common consumer goods. Observing the production of silver-plated tableware can shed light on how this process works.

Figure 14.14: Electroplating with copper.

Source: hBy Swoolverton - Template:Innoquick Electronics Limited (www.iqpcb.com), CC BY-SA 3.0, https://commons.wikimedia.org/w/index.php?curid=9905722

14.18 METALLURGY

Non-iron-and-steel-related technologically significant metals such as aluminum and titanium are obtained via electrochemical processes such as electrolysis in copper sulfate solutions, and copper is refined by electrolysis in solutions of copper and sulfuric acid.

14.19 PRODUCTION OF CHEMICALS

Currently, brine electrolysis to produce chlorine and caustic soda is one of the chemical industry's most significant volume producers. For example, today's processes can be used to make a wide range of inorganic compounds and synthetic fibres like nylon. In light of the possibility that future development of controlled fusion will significantly reduce electricity costs, intensive research in organic electrochemistry promises significant advancements in its application (Metz et al., 2014).

Inquiry into chemistry

Potentiometry, which measures the ionization constant using an electrode process, is an essential part of modern automated analytical chemistry.

14.20 INVESTIGATIONS INTO THE NATURAL WORLD

The idea that many biological processes, such as blood clotting and nerve impulse transmission, are electrochemical is spreading in biology. Electrochemical mechanisms are required to explain the efficient conversion of chemical energy from food into mechanical energy in the body. In a variety of directions, bioelectrochemistry is a hotly debated topic of study.

Summary

Electrochemistry is a branch of chemistry that studies how chemical and electrical energy can be converted. There are galvanic cells that generate an electric current in batteries. Batteries of the primary and secondary types are the most commonly used. In contrast to secondary batteries, primary batteries are "one-time-use" and cannot be recharged. Dry cells and alkaline cells make up the majority of primary batteries. Secondary batteries, which are rechargeable, are the second type of battery. Nickel-cadmium (NiCd), lead-acid, and lithium-ion cells are all examples of secondary batteries (Congrains et al., 2013). However, unlike batteries, fuel cells need to be replenished regularly with fuel and oxidizer. The hydrogen fuel cell produces water by combining hydrogen and oxygen from the air, making it more efficient than internal combustion engines. Electrolysis is a nonspontaneous reaction induced by the application of an external voltage.

CHAPTER 15

NUCLEAR CHEMISTRY

CONTENTS

Nuclear chemistry is part of the branches of chemistry. It deals with radioactivity, nuclear processes, and transformation in the nuclei of atoms. Some of the transformations in the nucleus of atoms include nuclear properties and nuclear transmutation. Nuclear chemistry can also be defined as the chemistry of radioactive elements such as radon, radium and actinides together with the chemistry associated with equipment that are designed to perform nuclear processes. Some of the equipment are nuclear reactors. It includes the corrosion of surfaces and behavior under conditions of normal and abnormal operation such as accidents (Mendoza & Núñez, 2009). One of the important area in nuclear chemistry is the behavior of the object and materials after being placed in the Buckhead waste storage and disposal site.

Nuclear chemistry includes the study of chemical effects that result from the absorption of radiation within plants, animals and other materials. The study of radiation in chemistry is referred to as radio chemistry and it controls much of radiation biology as radiation is able to affect living things at the molecular scale. It can also be explained by radiation altering biochemicals within an organism. After the alteration of the biomolecules there is a change in the chemistry that occurs within the organism. The change in chemistry leads to a biological outcome. For this reason, nuclear chemistry is said to greatly assist with the understanding of medical treatments including cancer radiotherapy. It also enables the improvement of such treatment.

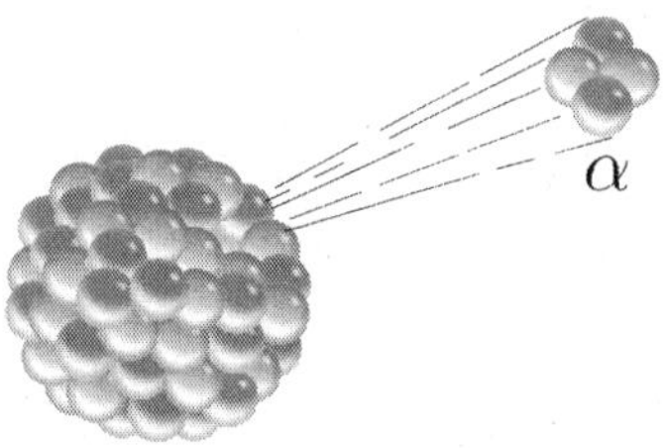

Figure 15.1: Nuclear chemistry involves the study of nuclear processes.

Source: By Inductiveload - Own work, Public Domain, https://commons.wikimedia.org/w/index.php?curid=2858666

Nuclear chemistry also involves the study of the production and use of radioactive sources for different kinds of processes. Some of these processes include radiotherapy in medical applications, utilization of radioactive tracers within industry, environment and science and the utilization of radiation in modifying materials such as polymers. Nuclear chemistry covers the study and use if nuclear processes in non-radioactive areas of human activity. For example, nuclear magnetic resonance spectroscopy that is used in synthetic organic chemistry and physical chemistry and in structural analysis in macro-molecular chemistry (Campos & Vasconcelos, 2010). Nuclear chemistry is at times referred to as radiochemistry,it includes the study of the formation of elements in the universe, designing radioactive drugs for diagnostic Medicine and other applications in other fields.

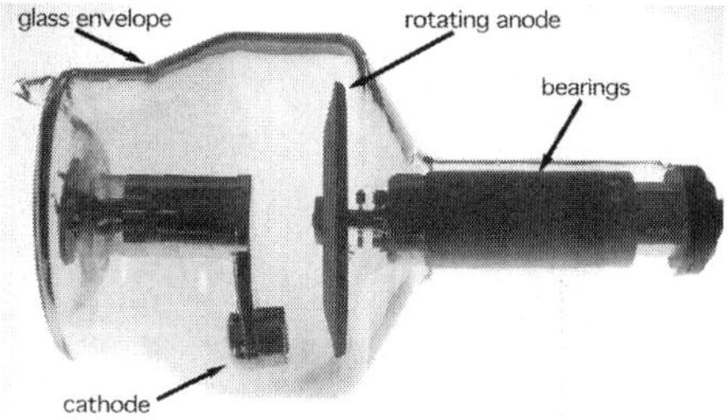

Figure 15.2: Production of X-rays is covered in nuclear chemistry.

Source: By Daniel W. Rickey - Own work, CC BY-SA 3.0, https://commons.wikimedia.org/w/index.php?curid=16622562

15.1 NUCLEAR REACTIONS AND RADIATIONS

Figure 15.3: Rutherford played an important role in the classification of radiations.

Source: By Bain News Service, publisherRestored by: Bammesk - Library of CongressCatalog: https://lccn.loc.gov/2014716719Image download: https://

cdn.loc.gov/service/pnp/ggbain/36500/36570v.jpgOriginal url: https://www.loc.gov/pictures/item/2014716719/, Public Domain, https://commons.wikimedia.org/w/index.php?curid=112190894

When studying nuclear chemistry, topics such as nuclear reactions and radiations will not be avoided. There are three types of radioactive radiation. These radiations were sorted out by Rutherford in 1902. He was able to sort out the radiations by passing discharge between two oppositely charged plates. Some of the radiations bent towards the negative plates and carry positive charges were known as alpha rays while the ones that bent towards the positive plate carried negative charges also known as beta rays. The other category of radiations are those said to be uncharged as they passed straight through the electric field and are also known as gamma rays (Mansour et al., 2010). The different types of radiation are formed differently. Unlike the normal chemical reactions that form molecules, in nuclear reactions, there is the transformation of an element from one form to another. The transformation property is used in nuclear power plants when producing nuclear energy.

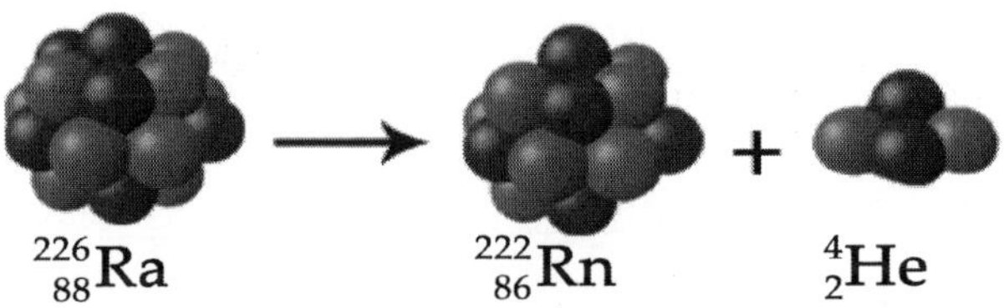

The nucleus of an atom of radium-226 contains 88 protons and 138 neutrons. A radium-226 nucleus undergoes alpha decay to form a different element, radon-222, and an alpha particle.

Figure 15.4: An alpha particle is usually represented by the Helium molecule.

Source: https://upload.wikimedia.org/wikipedia/commons/a/a1/Alpha-decay.png

Alpha radiation is the emission of an alpha particle in the atom's nucleus. The alpha particle is made up of two protons and two neutrons. The alpha particle is very similar to the Helium nucleus. When an alpha particle is emitted by an atom, the atomic mass of an atom decreases by 4 units. Beta radiation on the other hand involves the transmutation of a neutron into an electron and neutron. When a beta particle is emitted by an atom, the mass of the atom does not change. However, the atomic number increases by one. Gamma radiation involves the emission of electromagnetic energy from the nucleus of the atom. During gamma radiation, no particles are emitted. This means that there is no transmutation of atoms. Some nuclear reactions are stimulated artificially. In most cases, elements undergo natural radioactive decay (Colussi et al., 2013). Some simulated nuclear reactions include nuclear fission which is a type of reaction where the atom's nucleus splits into smaller parts resulting in the release of large amounts of energy in the process. The other reaction is nuclear fusion which involves two or more elements being fused together resulting in the formation of a larger element that releases a large amount of energy in the process.

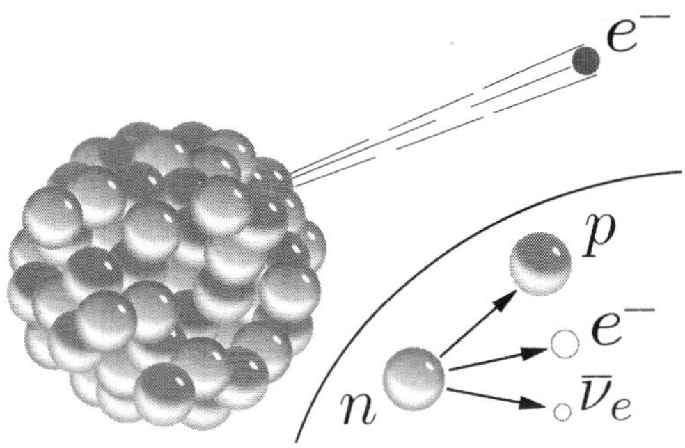

Figure 15.5: Beta radiation involves the transmutation of a neutron to an electron.

Source: https://en.wikipedia.org/wiki/File:Beta-minus_Decay.svg

15.2 COMPONENTS OF THE NUCLEUS

Nuclear chemistry revolves around the composition of the nucleus. Knowledge of the elements of the nucleus is to better understand the factors that affect nuclear stability. In most cases, known elements are said to have at least one isotope whose atomic nucleus is stable indefinitely. It is also a known fact that all elements have isotopes that disintegrate as they are unstable. These isotopes undergo disintegration and decay at measurable rate through the emission of radiation. Here are elements that do not have stable isotopes and decay to other elements. The process of nuclear decay leads to changes in the manner in which valence electrons of atoms are arranged. Nuclear decay leads to changes in the atomic nucleus.

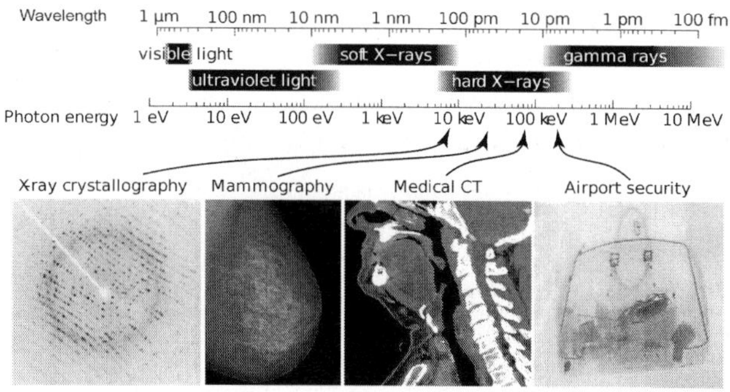

Figure 15.6: The composition of the nucleus gives more information on the formation of X-rays.

Source: By Ulflund - This figure is a compilation of different images from wikimedia commons.The graph at the top I have made myself, originaly uploaded as .Own workThe crystallography image is from File:Lysozym diffraction.png by user:Del45.The mammography image is from File:40F MLO DMMG.png by Nevit Dilmen (talk).The CT image is from File:Ct-workstation-neck.jpg by en:User:ChumpusRex.The luggage scanner image is from File:Luggage screening at VTBS.JPG by user User:Mattes., CC BY-SA 3.0, https://commons.wikimedia.org/w/index.php?curid=22545004

Elements of the nucleus include the atomic nucleus. Usually, each element can be represented by the notation $^{X}A_{Z}$ where the letter A is used to refer to the mass number which is the sum of the number of neutrons and the number of protons while the letter Z is used to represent the atomic number which is the number of protons. The nucleus of an atom is made up of neutrons and protons. Collectively, the neutron and protons are known as nucleons. The term nuclide is used to refer to an atom with a particular number of neutrons and protons. Nuclides having the same number of protons but different numbers of neutrons are known as an isotope (Marian, 2012). Representation of isotopes involves the use of an alternative notation that uses the band of the element followed by the mass number. For example with carbon, it becomes carbon-12. Oxygen has different isotopes that can be written as oxygen-16, oxygen-17 and oxygen-18. Usually, the number of neutrons is equal to A-Z. It is noted that the first isotope of oxygen has 8 neutrons, the second isotope had 9 neutrons while the third isotope has 10 neutrons. In most cases, isotopes of naturally occurring elements are found in nearly fixed proportions whereby each proportion constitutes an isotope of natural abundance. This is well demonstrated by terrestrial samples of oxygen where about 99.76% of the oxygen atom is oxygen-16, 0.29% is oxygen-28 and the remaining 0.94% is oxygen-7. If the nucleus is unstable or undergoes spontaneous decay is then said to be radioactive. It leads to the emission of subatomic particles and electromagnetic radiation. The resultant emission is collectively referred to as radioactivity. The level of radioactivity of a metal can be measured. The term radioisotope is used to refer to isotopes with the ability to emit radiation. Chemical kinetics is used to describe and allow one to understand the rate at which radioactive decay occurs. The rate of radioactive decay is characteristic of the isotopes. Radioactive decay is reported as half-time of the amount of time required for half of the initial number of nuclei present to decay in a first-order reaction (Chen et al., 2015). The half-life of isotopes usually ranges from fractions of a second to billions of years and among other applications. The half-life of objects is used in measuring the age of ancient objects. Calculations are done on the approximated half-life of the object and the results are used to estimate the age of the object.

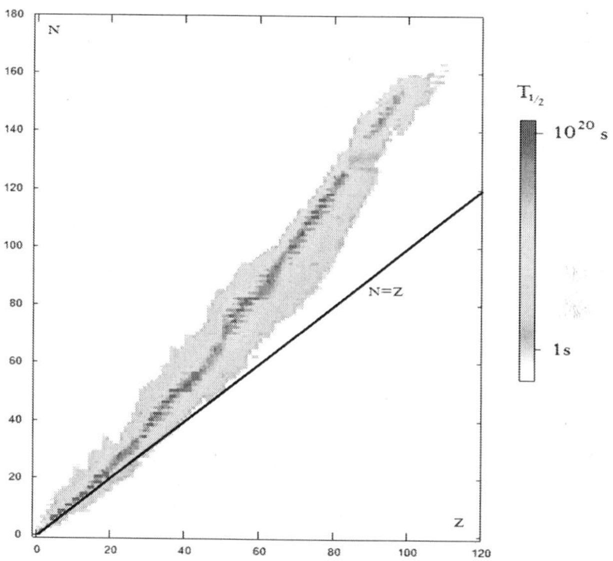

Figure 15.7: Different elements have different half-life.

Source: https://upload.wikimedia.org/wikipedia/commons/7/7a/Isotopes_and_half-life.PNG

15.3 NUCLEAR STABILITY

The nucleus of an atom is said to occupy a tiny fraction of the volume of an atom. The atom usually contains a number of neutrons and protons which is characteristic of given isotopes. In an atom, the positively charged protons repel each other through electrostatic repulsion. However, the nucleus does not fly apart. This is attributed to the presence of a strong nuclear force. The strong nuclear force is the extremely powerful but short-range attractive force present between nucleons. This means that there are competing interactions within the atomic nucleus. In most cases, a stable nuclei in the exception of the hydrogen-1 nucleus are said to contain at least one neutron with the ability to overcome the electrostatic repulsion present between protons (Chaplin, 2006). Any increase in the number of protons in the nucleus causes the need for the rapid increase in the number of neutrons in the nucleus. If there are several protons in the nucleus but very few neutrons, then there will be an imbalance of forces in the nucleus leading to nuclear instability.

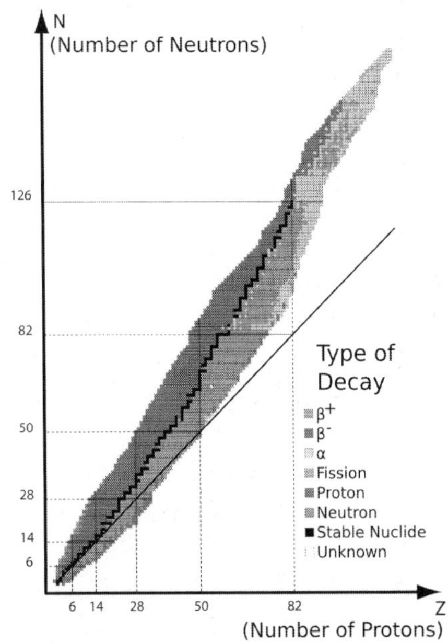

Figure 15.8: The stability of an element will determine whether it will undergo the decay process to produce stable elements.

Source: By Table_isotopes.svg: Napy1kenobiderivative work: Sjlegg (talk) - Table_isotopes.svg, CC BY-SA 3.0, https://commons.wikimedia.org/w/index.php?curid=6703703

The presence of electrostatic repulsions between positively charged protons would normally lead to the nucleus of atoms flying apart. For stable atomic nuclei, the repulsions are overcome by the strong nuclear force. The present relationship between the number of neutrons and the number of protons in stable nuclei is usually defined as having half-life that is longer than 10 times the age of the planet earth. 1He and 3He are the only two stable isotopes whose neutron to proton ratio is less than 1. There are several stable isotopes of light atoms with a neutron to proton ratio equivalent to one (Chemat et al., 2012). All other stable nuclei have a higher neutron to proton ratio that may steadily increase to about 1.5 for the heaviest nuclei. Despite the number of neutrons, elements whose Z is greater than 83 are said to be radioactive and unstable. Elements with the magic number of nucleons are said to demonstrate stability.

The demonstrated patterns are reminiscent of the stability associated with the closed-shell electron configuration of noble gases found in group 18. This led to the hypothesis that the nucleus has shells of nucleons that are analogous to the shells occupied by the electrons in the atom. A graph showing the relationship between the neutron to proton ratio and nuclear stability shows that the peninsula of stable isotopes that are surrounded by a reef of radioactive isotopes. These radioactive isotopes are said to have a stability that allows them to exist for varying lengths of time. This means that they eventually decay to produce other nuclei.

When discussing nuclear chemistry, the term super-heavy elements comes up. The peninsula of stability is used to show the stability of various kinds of elements. It is also noted that the relationship between the neutron to proton ratio and nuclear stability shows a small island of stability that is said to exist in the upper right corner. The island is said to correspond to the super-heavy elements. Such elements have atomic numbers near the magic number 126. It is expected that the next magic number of neutrons should be 184. For this reason, it is suggested that elements with 184 neutrons and 114 protons might be stable enough to exist in nature. Most of these claims were met with skepticism for a number of years. Some of the atoms of isotopes that faces skepticism include those with Z=114 and Z=116. These isotopes have faced skepticism since 1999 as they were prepared and found to be surprisingly stable. For example, the isotope with Z= 144 lasts 2.7 seconds before decaying and was therefore classified as an eternity by nuclear chemists (Matranga & Corsi, 2012). More evidence has been generated to show that there is the inexistence of a nucleus with A= 292. This was

Thorium-232. The element thorium was found to have an estimated half-life that is greater than 10^8 years. This means that it is practically stable. Measurements of its mass were consistent with predictions for the mass of an isotope with Z=122. The super-heavy elements are those with a number of relatively long-lived nuclei.

15.4 NUCLEAR REACTIONS

Knowledge of nuclear reactions is useful in knowing the different kinds of radioactive decay and properly balance a nuclear reaction. There are two genera kinds of nuclear reactions which are nuclear transmutation and nuclear decay reactions. Nuclear decay reactions are also known as radioactive decay. This kind of reaction involves an unstable nucleus emitting radiation and is transformed into the nucleus of one or more other elements. The generated daughter nuclei have a lower mass and lower energy and is said to be more stable than the parent nucleus that decayed. On the other hand, nuclear transmutation reactions involve the nucleus reacting with a subatomic particle or another nucleus forming a product nucleus that is more massive compared to the starting material (Lovley, 2003). In most cases, nuclear reactions tend to occur spontaneously under all conditions which are very different from nuclear transmutation reactions occur only under very special conditions. This includes the collision of a bean of highly energetic particles with a target nucleus. It is important to note that there are different classes of radioactive nuclei along with their characteristic nuclear decay reactions and the emitted radiation. A notable pattern is that nuclear decay reactions occur spontaneously within all conditions while nuclear transmutation reactions are induced.

There are different classes of radioactive nuclei; three to be precise all of which are characterized by a different decay process or set of processes. The first class includes the neutron-rich nuclei. This involves the nuclei on the upper left side of the band of stable nuclei in the relationship between neutron-to-proton and nuclear stability ratio having a neutron-to-proton ratio that is too high to give a stable nucleus. This kind of nuclear decay involves the conversion of a neutron to a proton that leads to a decline in the neutron-to-proton ratio. The second classification covers the neutron-poor nuclei whereby the nuclei on the lower right side of the band of stable nuclei have a neutron-to-proton ratio that is very low to give a stable nucleus. In such a nuclei decay processare known to have a net effect of converting a proton to a neutron leading to an increase in the neutron-to-proton ratio.

The third class involves heavy nuclei. This class has very few exceptions. It includes heavy nuclei whose A is greater or equal to 200 and is intrinsically unstable despite the neutron-to-proton ratio. It also has all nuclei with Z greater than 83 being unstable. This is attributed to the cumulative effects of electrostatic repulsions between the large number of positively charged protons that cannot be totally overcome by string nuclear force despite the number of neutrons present.

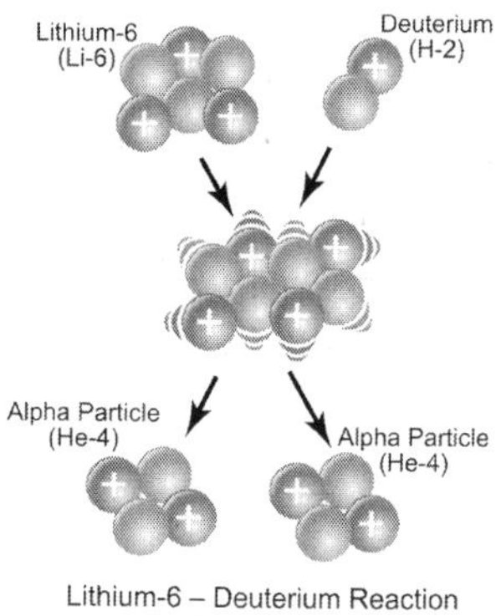

Figure 15.9: Some nuclear reactions are artificially caused.

Source: By Sakurambo - Own work, based on Image:Li6-D Reaction.png, CC BY-SA 3.0, https://commons.wikimedia.org/w/index.php?curid=2096742

For such nuclei, it tends to decay through the emission of an alpha particle with a helium nucleus and decreases the number of neutrons s d proto s in the original nucleus by 2. This is attributed to the fact that the neutron-to-proton ratio in an alpha particle is 1. The net result of alpha emission is an increase in the neutron-to-proton ratio. An important point to note is that nuclear decay reactions lead to the production of daughter nuclei with a more favourable neutron-to-proton ratio that makes it more stable than the parent nucleus.

15.5 NUCLEAR DECAY REACTIONS

The number and type of atoms are used in various ways in nuclear chemistry. They are used in presenting balanced chemical equations and writing nuclear equations for a nuclear decay reaction. The procedure followed allows the prediction of the identity of either the parent or the daughter nucleus when the identity of one element is known. The total number of nucleons is conserved in all nuclear reactions despite the mode of decay. The $^{X}A_{Z}$ notation has been extended by chemists to better describe the nuclear decay reactions. There are different symbols used to represent the different forms of emitted radiation. The most common type of particles released from radiation includes beta and alpha particles while gamma rays are the most common types of rays released from reactions (Lischka et al., 2011). In chemical kinetics, the most notable addition is the positron which is a particle having the same mass as an electron but a positive charge rather than the expected negative charge. There are six fundamental kinds of nuclear decay reactions all of which release different kinds of particles of energy. Among the common modes of nuclear decay include beta and alpha decay and gamma emission. The others are key to understanding the other forms of nuclear decay reactions.

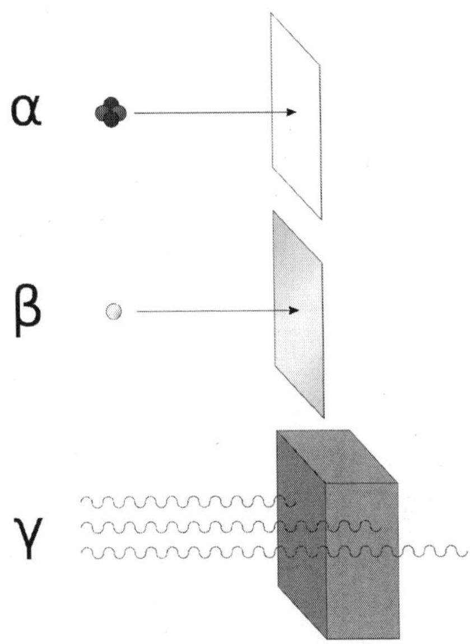

Figure 15.10: In most nuclear reactions gamma rays are produced in the form of energy.

Source: By Stannered - Traced from this PNG image., CC BY 2.5, https://commons.wikimedia.org/w/index.php?curid=1716989

Alpha decay is common in many nuclear reactions. It commonly occurs in nuclei with mass numbers greater than 200. This means that there is the emission of a helium-4 nucleus as the alpha particle. The general reaction is as follows: $^{X}A_{Z}$parent→X′A−4Z−2daughter+α42alphaparticle

In the reaction, the daughter nuclide contains two fewer neutrons and two fewer protons than the parent. This means that the alpha particle emission results in the production of a daughter nucleus with a mass number A-4 and a nuclear charge Z-2 compared to the parent nucleus. A good example is with Radium-226 huh under alpha decay forming radon-222. $^{226}Ra_{88}a \rightarrow ^{222}Ra_{86} + \alpha 42$. This means that nucleons are conserved in this and other nuclear reactions. The summation of the mass of numbers of the products, 22+4=226, which is the mass number of the parent. In the same manner, the sum of the atomic number of the products, 86 +2= 88which is the atomic number of the parent therefore the nuclear equation is balanced (Dubernet et al., 2010). In beta decay, it is common in nuclei where the neutrons undergo beta decay that involves the conversion of the neutron to a proton and a high-energy electron is ejected from the nucleus as beta particle. The general reaction for the beta decay is $^{X}A_{Z}$parent$\rightarrow ^{X'}A_{Z}$+1 daughter+β0−1beta particle. In this kind of reaction, there is no change in the mass number of the nucleus. It however results in the increase of +1 in the atomic number attributed to the addition of protons in the daughter nucleus. For this reason, the beta decay results in the decrease in the neutron-to-proton ratio that makes the nucleus moves towards the band of stable nuclei. A good illustration is with carbon-14 that exhibits beta decay forming nitrogen -14. In this case, the number of nucleons is conserved, and the charges are balanced. Both the parent and daughter have the same mass number,14, and the sum of the atomic numbers of the products is 6 which is similar to the atomic number of the carbon-14 parent.

In positron emission, it involves a positron has the same mass as an electron but opposite charges. For this reason, the position emission is very different from the beta decay. Positron emission is said to be characteristic of neutron-poor nuclei. This kind of decay occurs through the transformation of a proton to a neutron and the emission of high-energy positron. Positron decay bears similarities to beta decay in that there is no change in the mass number of the nucleus. The difference is noted in the atomic number of the daughter nucleus which is lower by 1 than that of the parent. This means that there is an increase in the neutron-to-proton ratio that moves the nucleus close to the band of stable nuclei. A good example is carbon-11 which experiences positron emission resulting in the formation of boron-11. As noted from the equation, the nucleons are conserved and there is a balancing of charges. There is no change in the mass number,11, and the sum of the atomic numbers of the product is 6 similar to the atomic number of the parent carbon-11 nuclide. There is electron capture that involves a neutron-poor nucleus decaying either through electron capture or positron emission (Lisio et al., 2019). It involves an electron in the inner shell reacting with a proton thereby producing a neutron. P11+e0−1→n10 is the equation for the reaction An X-ray is emitted when a second electron moves from an outer shell to take the place of the lower-energy electron that is absorbed by the nucleus. The general equation for the electron capture reaction is as follows: ${}^{X}A_{Z}$parent+e0−1electron→${}^{X'}A_{Z}$−1daughter+x-ray.

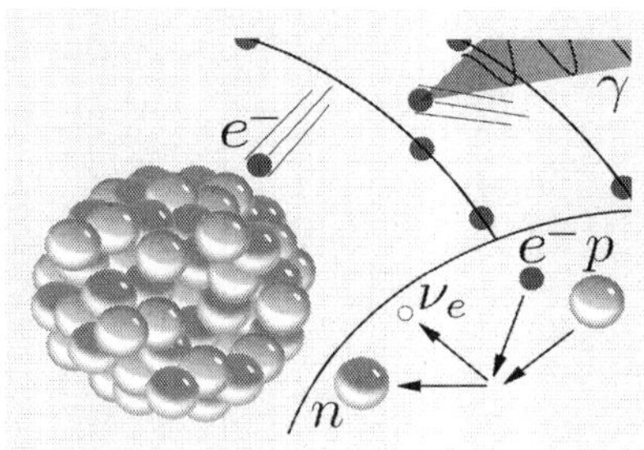

Figure 15.11: Electron capture involves an electron being removed and being used in the production of a neutron.

Source: https://commons.wikimedia.org/wiki/File:Electron_capture.svg

In the electron capture, there is no change in the mass number of the nucleus as the formed neutron and the lost proton have a mass number of 1. Similar to positron emission, the atomic number of the daughter nucleus is lower by 1 compared to that of the parent. This means that there is an increase in the neutron-to-proton ratio that moves the nucleus towards the band of stable nuclei. This is well illustrated by iron-55 that decays through electronic capture forming manganese-55 and the equation is written as follows: ${}^{55}Fe_{26}$e+e0^{-1}→M${}^{55}Mn_{25}$n+x-ray. It is noted that there is a deviation in the atomic number of parent and daughter nuclides though there is a similarity in the mass number. When writing a balanced nuclear equation for such kind of a reaction, it is very necessary to include the captured electron in the equation as illustrated ${}^{55}Fe_{26}$e+e0^{-1}→M${}^{55}Mn_{25}$n+x-ray. In electron capture and positron emission, such kind of reactions are noted for nuclides with low neutron-to-proton ratio though the decay rates for the two processes may be different (Dufrêne et al., 2017).

In gamma emission, there are nuclear decay reactions that produce daughter nuclei that occur in nuclear excited states that are similar to an atom in which an electron is excited to a higher-energy orbital into a given electronic excited state. An electron in the electronic excited state will emit energy in the form of a photon as it returns to the ground state. The same happens to a nucleus in an excited state that releases energy in the form of a photon as it returns to the ground state. The released high energy photons are the gamma rays. There are cases where the gamma emission is virtually instantaneous. This is noted in the alpha decay of uranium-238 that decays to thorium-234 as is noted in the following equation:

$U^{238}U_{92}$→$T^{234}Th_{90}$h*excitednuclearstate+α42relaxation→T^{234}Th90h +γ00. If the decay event that created the excited nucleus is disregarded, then the equation becomes $^{234}Th_{88}$h*→$T^{234}Th_{88}$hh+γ00. It is common for gamma emission to occur after a significant delay. This is the case for technetium-99m with a half-life of about 6 hours before it emits gamma rays from technetium-99. When the letter m is used beside the element, it is used to indicate that it is metastable. Gamma rays are emitted as energy meaning that their emission does not affect the mass number of the daughter nuclide or its atomic number. For this reason, gamma-ray emission is the only kind of radiation that does not necessarily involve the conversion of one element to another. In most cases, it is observed in conjunction with some other nuclear decay reaction.

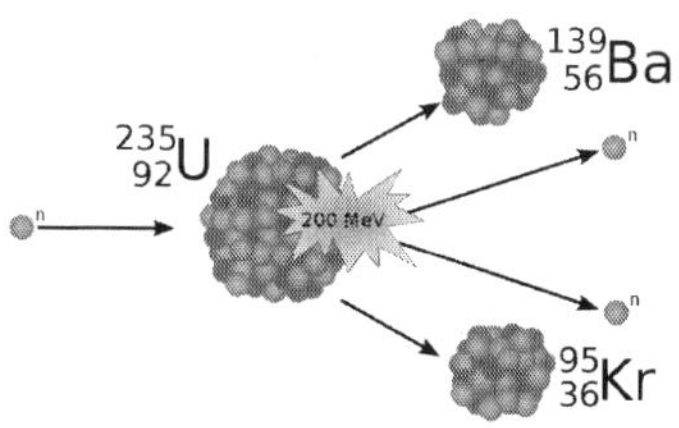

Figure 15.12: Spontaneous fission mostly occurs in massive nuclei.

Source: By Original: Stefan-Xp (talk · contribs) / Vectorization: Wondigoma (talk · contribs) - This file was derived from: Kernspaltung.png:, CC BY-SA 3.0, https://commons.wikimedia.org/w/index.php?curid=8540436

There is spontaneous fission that occurs in massive nuclei with high neutron-to-proton ratios. This process results in the nucleus being broken down into pieces that have spontaneous fission whereby the nucleus is broken down into two pieces that have different atomic masses and atomic numbers. The process of spontaneous fission is considered the most important for the transactinide elements whose Z is greater or equal to 103. Usually, spontaneous fission is usually accompanied by the emission of several neutrons (Laganà et al., 2019). This is the case in the spontaneous fission of C25498f. It gives a distribution of fission products and one possible set of products is alpha decay, electron capture and beta decay

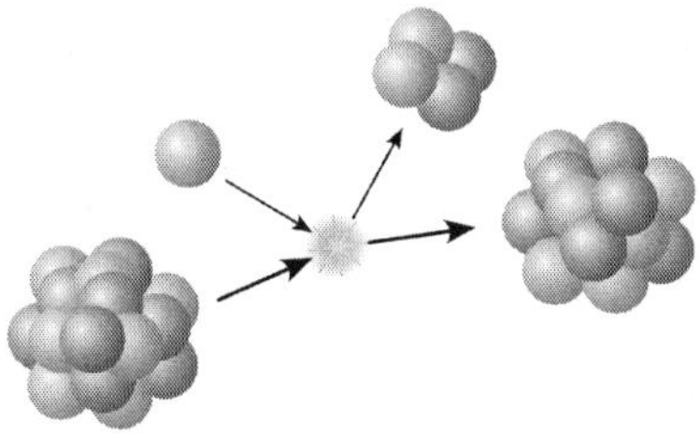

Figure 15.13: Spontaneous fission leads to the emission of several neutrons.

Source: By Kjerish - This file has been extracted from another file, CC BY-SA 4.0, https://commons.wikimedia.org/w/index.php?curid=54378478

15.6 THE RADIOACTIVE DECAY SERIES

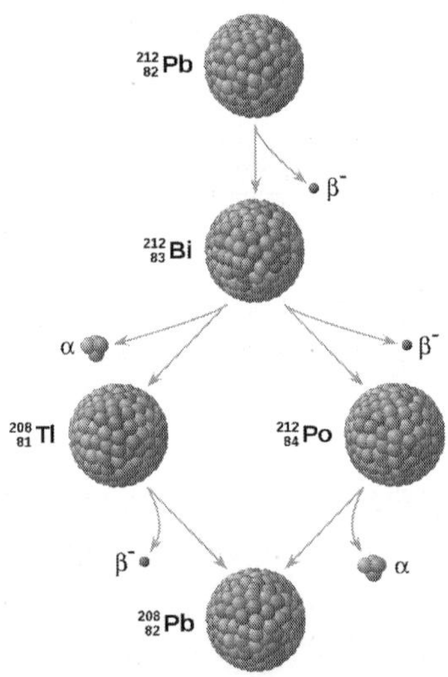

Figure 15.14: Radioactive decay series explains the presence of some elements on earth.

Source: By Eugene Alvin Villar (seav) - Own work, CC BY-SA 4.0, https://commons.wikimedia.org/w/index.php?curid=4627839

A great part of nuclear chemistry looks into the radioactive decay series. It is common knowledge that nuclei of all elements with atomic numbers greater than 83 are unstableness. This means that isotopes of elements beyond bismuth in the periodic table are radioactive. Usually, alpha decay caused a decrease in Z by only 2 while the electron capture or positron capture causes a decrease in Z by 1. This makes it impossible for any nuclide whose Z is greater than 85 to decay to a stable daughter nuclide in a single step with exception of nuclear fission. This means that radioactive isotopes with Z greater than 85 usually decay to a daughter nucleus until it attains a stable nucleus is attained. The series of sequential alpha- and beta- decay is referred to as the radioactive decay series. The uranium-238 decay series is the most common radioactive decay series. This series results in the production of lead-206 in a series of 14 sequential alpha- and beta- decay reactions (Ebbesen, 2016). The radioactive decay series can be written for almost any isotope whose Z is greater than 85, there are only two that occur naturally. They include the decay of uranium-235 to lead-207 and thorium-232 to lead-297. The fourth series in the decay of neptunium-237 to bismuth-208 in 11 steps. This series is said to have occurred on the primitive earth. Neptunium had a half-life of only 2.14 million years. All the neptunium-236 found on earth was formed long ago through the decay. Currently, all the neptunium on the Earth's surface is artificial or synthetic.

15.7 INTERACTION OF NUCLEAR RADIATION WITH MATTER

Usually, nuclear reactions do not affect the valence electron of the atom in question. This is in exception of electron capture where an electric charge draws an electron from an orbital having the lowest energy level. Generally, nuclear reactions do not directly cause chemical changes. However, the particles and photons emitted during the decay process have high energies and they can indirectly produce chemical changes in the matter that surrounds the nucleus that has decayed. This is well explained by an alpha particle which is an ionized helium nucleus with the ability to act as a powerful oxidant. Emitted radiation has the ability to describe how radiation is able to interact with matter. From this, it was able to classify radiation as ionizing or non-ionizing.

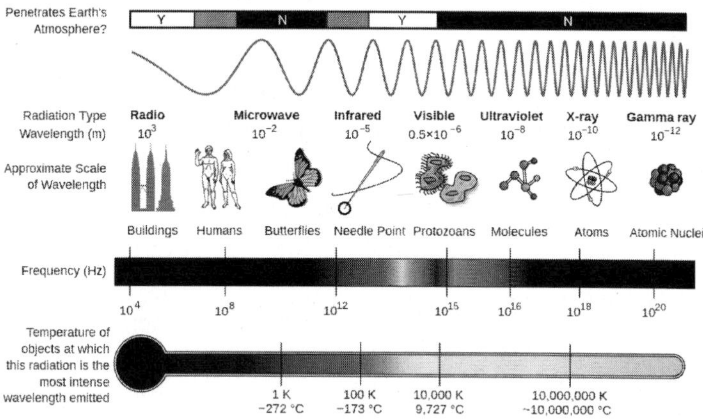

Figure 15.15: There are chemical and biological effects of radiation.

Source: CC BY-SA 3.0, https://commons.wikimedia.org/w/index.php?curid=2974242

Ionizing radiation usually affects the body of human beings after exposure (Kleber &Johnson, 2010). Such kind of radiation has the ability to raise the stability of an ion. Exposure to radiation results in both chemical and biological effects of radiation. Further studies in nuclear chemistry clearly define the natural and artificial sources of radiation.

REFERENCES

1. Antoni, D., Burckel, H., Josset, E., & Noel, G. (2015). Three-dimensional cell culture: a breakthrough in vivo. *International journal of molecular sciences*, *16*(3), 5517-5527.
2. Aramwit, P., Kanokpanont, S., Nakpheng, T., & Srichana, T. (2010). The effect of sericin from various extraction methods on cell viability and collagen production. *International Journal of Molecular Sciences*, *11*(5), 2200-2211.
3. Aidas, K., Angeli, C., Bak, K. L., Bakken, V., Bast, R., Boman, L.,& Ågren, H. (2014). The D alton quantum chemistry program system. *Wiley Interdisciplinary Reviews: Computational Molecular Science*, *4*(3), 269-284.
4. Arif, I. A., Bakir, M. A., Khan, H. A., Al Farhan, A. H., Al Homaidan, A. A., Bahkali, A. H., ... &Shobrak, M. (2010). A brief review of molecular techniques to assess plant diversity. *International Journal of Molecular Sciences*, *11*(5), 2079-2096.
5. Barbatti, M., Ruckenbauer, M., Plasser, F., Pittner, J., Granucci, G., Persico, M., & Lischka, H. (2014). Newton-X: a surface-hopping program for nonadiabatic molecular dynamics. *Wiley Interdisciplinary Reviews: Computational Molecular Science*, *4*(1), 26-33.
6. Bacellar, I. O., Tsubone, T. M., Pavani, C., & Baptista, M. S. (2015). Photodynamic efficiency: from molecular photochemistry to cell death. *International journal of molecular sciences*, *16*(9), 20523-20559.

7. Bennett, N. C., Gardiner, R. A., Hooper, J. D., Johnson, D. W., &Gobe, G. C. (2010). Molecular cell biology of androgen receptor signalling. *The international journal of biochemistry & cell biology*, *42*(6), 813-827.
8. Brownlee, M. (2001). Biochemistry and molecular cell biology of diabetic complications. *Nature*, *414*(6865), 813-820.
9. Congrains, A., Kamide, K., Ohishi, M., &Rakugi, H. (2013). ANRIL: molecular mechanisms and implications in human health. *International journal of molecular sciences*, *14*(1), 1278-1292.
10. Campos, A., & Vasconcelos, V. (2010). Molecular mechanisms of microcystin toxicity in animal cells. *International journal of molecular sciences*, *11*(1), 268-287.
11. Colussi, D., Brandi, G., Bazzoli, F., & Ricciardiello, L. (2013). Molecular pathways involved in colorectal cancer: implications for disease behavior and prevention. *International journal of molecular sciences*, *14*(8), 16365-16385.
12. Chen, Y., Zou, C., Mastalerz, M., Hu, S., Gasaway, C., & Tao, X. (2015). Applications of micro-fourier transform infrared spectroscopy (FTIR) in the geological sciences—a review. *International journal of molecular sciences*, *16*(12), 30223-30250.
13. Chaplin, M. (2006). Do we underestimate the importance of water in cell biology?. *Nature Reviews Molecular Cell Biology*, *7*(11), 861-866.
14. Chemat, F., Vian, M. A., &Cravotto, G. (2012). Green extraction of natural products: Concept and principles. *International journal of molecular sciences*, *13*(7), 8615-8627.
15. Dubernet, M. L., Boudon, V., Culhane, J. L., Dimitrijevic, M. S., Fazliev, A. Z., Joblin, C., ... &Zeippen, C. J. (2010). Virtual atomic and molecular data centre. *Journal of Quantitative Spectroscopy and Radiative Transfer*, *111*(15), 2151-2159.
16. Dufrêne, Y. F., Ando, T., Garcia, R., Alsteens, D., Martinez-Martin, D., Engel, A., & Müller, D. J. (2017). Imaging modes of atomic force microscopy for application in molecular and cell biology. *Nature nanotechnology*, *12*(4), 295-307.
17. Ebbesen, T. W. (2016). Hybrid light–matter states in a molecular and material science perspective. *Accounts of chemical research*, *49*(11), 2403-2412.

18. Gentile, P., Chiono, V., Carmagnola, I., & Hatton, P. V. (2014). An overview of poly (lactic-co-glycolic) acid (PLGA)-based biomaterials for bone tissue engineering. *International journal of molecular sciences*, *15*(3), 3640-3659.
19. Haidekker, M. A., & Theodorakis, E. A. (2010). Environment-sensitive behavior of fluorescent molecular rotors. *Journal of biological engineering*, *4*(1), 1-14.
20. Hasanuzzaman, M., Nahar, K., Alam, M. M., Roychowdhury, R., & Fujita, M. (2013). Physiological, biochemical, and molecular mechanisms of heat stress tolerance in plants. *International journal of molecular sciences*, *14*(5), 9643-9684.
21. Huang, S. Y., & Zou, X. (2010). Advances and challenges in protein-ligand docking. *International journal of molecular sciences*, *11*(8), 3016-3034.
22. Kalra, H., Drummen, G. P., & Mathivanan, S. (2016). Focus on extracellular vesicles: introducing the next small big thing. *International journal of molecular sciences*, *17*(2), 170.
23. Karsenti, E. (2008). Self-organization in cell biology: a brief history. *Nature reviews Molecular cell biology*, *9*(3), 255-262.
24. Kawser Hossain, M., Abdal Dayem, A., Han, J., Yin, Y., Kim, K., Kumar Saha, S., ... & Cho, S. G. (2016). Molecular mechanisms of the anti-obesity and anti-diabetic properties of flavonoids. *International journal of molecular sciences*, *17*(4), 569.
25. Kharat, M., Du, Z., Zhang, G., & McClements, D. J. (2017). Physical and chemical stability of curcumin in aqueous solutions and emulsions: Impact of pH, temperature, and molecular environment. *Journal of agricultural and food chemistry*, *65*(8), 1525-1532.
26. Krylov, A. I., & Gill, P. M. (2013). Q-Chem: an engine for innovation. *Wiley Interdisciplinary Reviews: Computational Molecular Science*, *3*(3), 317-326.
27. Kleber, M., & Johnson, M. G. (2010). Advances in understanding the molecular structure of soil organic matter: implications for interactions in the environment. *Advances in agronomy*, *106*, 77-142.
28. Laganà, A. S., Garzon, S., Götte, M., Viganò, P., Franchi, M., Ghezzi, F., & Martin, D. C. (2019). The pathogenesis of endometriosis: molecular and cell biology insights. *International journal of molecular sciences*, *20*(22), 5615.

29. Leri, M., Scuto, M., Ontario, M. L., Calabrese, V., Calabrese, E. J., Bucciantini, M., & Stefani, M. (2020). Healthy effects of plant polyphenols: molecular mechanisms. *International journal of molecular sciences*, *21*(4), 1250.
30. Lisio, M. A., Fu, L., Goyeneche, A., Gao, Z. H., & Telleria, C. (2019). High-grade serous ovarian cancer: basic sciences, clinical and therapeutic standpoints. *International journal of molecular sciences*, *20*(4), 952.
31. Lischka, H., Müller, T., Szalay, P. G., Shavitt, I., Pitzer, R. M., & Shepard, R. (2011). Columbus—a program system for advanced multireference theory calculations. *Wiley Interdisciplinary Reviews: Computational Molecular Science*, *1*(2), 191-199.
32. Lovley, D. R. (2003). Cleaning up with genomics: applying molecular biology to bioremediation. *Nature Reviews Microbiology*, *1*(1), 35-44.
33. Matranga, V., & Corsi, I. (2012). Toxic effects of engineered nanoparticles in the marine environment: model organisms and molecular approaches. *Marine environmental research*, *76*, 32-40.
34. Marian, C. M. (2012). Spin–orbit coupling and intersystem crossing in molecules. *Wiley Interdisciplinary Reviews: Computational Molecular Science*, *2*(2), 187-203.
35. Mansour, H. M., Sohn, M., Al-Ghananeem, A., & DeLuca, P. P. (2010). Materials for pharmaceutical dosage forms: molecular pharmaceutics and controlled release drug delivery aspects. *International journal of molecular sciences*, *11*(9), 3298-3322.
36. Mendoza, C., & Núñez, L. A. (2009). Virtual atomic and molecular data centre. In *Proceedings of the First EELA-2 Conference* (pp. 219-225). Madrid: CIEMAT.
37. Metz, S., Kästner, J., Sokol, A. A., Keal, T. W., & Sherwood, P. (2014). C hem S hell—a modular software package for QM/MM simulations. *Wiley Interdisciplinary Reviews: Computational Molecular Science*, *4*(2), 101-110.
38. Milo, R., Jorgensen, P., Moran, U., Weber, G., & Springer, M. (2010). BioNumbers—the database of key numbers in molecular and cell biology. *Nucleic acids research*, *38*(suppl_1), D750-D753.
39. Moseti, D., Regassa, A., & Kim, W. K. (2016). Molecular regulation of adipogenesis and potential anti-adipogenic bioactive molecules. *International journal of molecular sciences*, *17*(1), 124.

40. Nwodo, U. U., Green, E., & Okoh, A. I. (2012). Bacterial exopolysaccharides: functionality and prospects. *International journal of molecular sciences*, *13*(11), 14002-14015.

41. Neese, F. (2012). The ORCA program system. *Wiley Interdisciplinary Reviews: Computational Molecular Science*, *2*(1), 73-78.

42. O'Gorman, C., Lucas, R., & Taylor, B. (2012). Environmental risk factors for multiple sclerosis: a review with a focus on molecular mechanisms. *International journal of molecular sciences*, *13*(9), 11718-11752.

43. Ogino, S., Lochhead, P., Chan, A. T., Nishihara, R., Cho, E., Wolpin, B. M., ... & Giovannucci, E. (2013). Molecular pathological epidemiology of epigenetics: emerging integrative science to analyze environment, host, and disease. *Modern Pathology*, *26*(4), 465-484.

44. Patil, V., Tran, K. Q., &Giselrød, H. R. (2008). Towards sustainable production of biofuels from microalgae. *International journal of molecular sciences*, *9*(7), 1188-1195.

45. Pinzi, L., & Rastelli, G. (2019). Molecular docking: shifting paradigms in drug discovery. *International journal of molecular sciences*, *20*(18), 4331.

46. Pritchard, B. P., Altarawy, D., Didier, B., Gibson, T. D., & Windus, T. L. (2019). New basis set exchange: An open, up-to-date resource for the molecular sciences community. *Journal of chemical information and modeling*, *59*(11), 4814-4820.

47. Raies, A. B., & Bajic, V. B. (2016). In silico toxicology: computational methods for the prediction of chemical toxicity. *Wiley Interdisciplinary Reviews: Computational Molecular Science*, *6*(2), 147-172.

48. Rastogi, G., & Sani, R. K. (2011). Molecular techniques to assess microbial community structure, function, and dynamics in the environment. In *Microbes and microbial technology* (pp. 29-57). Springer, New York, NY.

49. Rhoads, D. D., Wolcott, R. D., Sun, Y., & Dowd, S. E. (2012). Comparison of culture and molecular identification of bacteria in chronic wounds. *International journal of molecular sciences*, *13*(3), 2535-2550.

50. Stuppia, L., Antonucci, I., Palka, G., & Gatta, V. (2012). Use of the MLPA assay in the molecular diagnosis of gene copy number alterations in human genetic diseases. *International journal of molecular sciences*, *13*(3), 3245-3276.

51. Sneskov, K., & Christiansen, O. (2012). Excited state coupled cluster methods. *Wiley Interdisciplinary Reviews: Computational Molecular Science*, *2*(4), 566-584.
52. Shi, J., Wood, W. B., Martin, J. M., Guild, N. A., Vicens, Q., & Knight, J. K. (2010). A diagnostic assessment for introductory molecular and cell biology. *CBE—Life Sciences Education*, *9*(4), 453-461.
53. Shin, J. W., Kwon, S. H., Choi, J. Y., Na, J. I., Huh, C. H., Choi, H. R., & Park, K. C. (2019). Molecular mechanisms of dermal aging and antiaging approaches. *International journal of molecular sciences*, *20*(9), 2126.
54. Thomford, N. E., Senthebane, D. A., Rowe, A., Munro, D., Seele, P., Maroyi, A., &Dzobo, K. (2018). Natural products for drug discovery in the 21st century: innovations for novel drug discovery. *International journal of molecular sciences*, *19*(6), 1578.
55. Turney, J. M., Simmonett, A. C., Parrish, R. M., Hohenstein, E. G., Evangelista, F. A., Fermann, J. T., ... & Crawford, T. D. (2012). Psi4: an open-source ab initio electronic structure program. *Wiley Interdisciplinary Reviews: Computational Molecular Science*, *2*(4), 556-565.
56. Vasapollo, G., Sole, R. D., Mergola, L., Lazzoi, M. R., Scardino, A., Scorrano, S., & Mele, G. (2011). Molecularly imprinted polymers: present and future prospective. *International journal of molecular sciences*, *12*(9), 5908-5945.

INDEX